Activation Functions

Yasin Kütük

Activation Functions

Activation Functions in Deep Learning with
LaTeX Applications

PETER LANG

**Bibliographic Information published by the
Deutsche Nationalbibliothek**

The Deutsche Nationalbibliothek lists this publication in the Deutsche Nationalbibliografie; detailed bibliographic data is available online at http://dnb.d-nb.de.

Library of Congress Cataloging-in-Publication Data

A CIP catalog record for this book has been applied for at the Library of Congress.

ISBN 978-3-631-87328-1 (Print)
E-ISBN 978-3-631-87670-1 (E-PDF)
E-ISBN 978-3-631-87671-8 (EPUB)
10.3726/b19631

© Peter Lang GmbH
Internationaler Verlag der Wissenschaften
Berlin 2022
All rights reserved.

Peter Lang – Berlin · Bern · Bruxelles · Istanbul · New York · Oxford · Warszawa · Wien

This publication has been peer reviewed.

www.peterlang.com

Contents

Contents

Introduction

In recent years, it can be easily observed that statistics has been renewing itself by means of recent advances in both the processing and the data. Processing technologies are not restricted to the CPU itself but encompasses recent developments in other technologies such as GPU, TPU and other unclassified processors as singular or parallel processing units. The data has also been growing day by day[1] with the emergence and widespread use of smartphone technologies which has accelerated, firstly, data creation, and then collection and its storage compared to other resources. These two components of the technology provide a great, convenient basis for the birth of Zeitgeist[2]: Artificial Intelligence (AI hereafter).

AI actually has a long history which can be traced back to the Mechanical Turk (Standage, 2002), a chess machine, controlled by a human who is hidden in the machine playing against other humans as an intelligent and fake-automated machine thinking moves. The breakthrough in AI was made by Alan Turing when trying to break and translate the codes used by axis forces in the Second World War, which made AI an academic discipline in the late 1950's. This breakthrough was based on categorical and logical implementation of codes, symbolic reasoning and mathematical expressions to imitate a human brain. The first neural network implementation called SNARC was carried out during those years McCorduck (1979). Russell et al. (1995) assert that AI is achieved by changing the way of thinking. It has been achieved by making computers more intelligent and learnable than just codes, which feeds computers consistently from the environment or feeds them with data created or provided by human. So, the utopia of aiming to create a machine thinking, acting and feeling like a human has become a reality thanks to the changes in the way of thinking; switching from creating categorical or logical algorithms that shape problem solving behaviour of computers to computers using information-based learning. Hence, the information, namely data, became the source of learning.

The human brain learns the information which is provided by itself or by the environment. In order to make machines intelligent, the first thing which should

---

1   https://www.forbes.com/sites/bernardmarr/2015/09/30/big-data-20-mind-boggling-facts-everyone-must-read/#7a16b1917b1e .
2   As frequently stated or referred, the concept of zeitgeist is attributed to Hegel (1861), but he used only "the spirit of the age" which is originated as "der Geist seiner Zeit" in German.

be done is to mimic the algorithm of learning. Previously, it was questioned if machines, after the problem was declared to them, can solve the problem if the problem exists and has been defined for the machine. The word is the key concept of the algorithms before the 21st century. After that, the learning could be built on whether the condition of "if" exists, as human brains do. So then, the machines are programmed to learn "if" statements' through their own capabilities, and the main condition for that is to feed into the machines the information, the data, and implicit programming. This process as a whole is called machine learning. Computers can learn from the data, they can make their own "if statements" by generating equation(s), establishing a relationship between informations, separating others etc. Machine learning is a discipline of learning from data in order to make predictions using them.

1. Machine Learning

The best and general definition of what machine learning may come from Mitchell (1997) explaining that "machine learning is a discipline of computer algorithms that improve automatically through experience". In the word experience, there is a hidden and a deep meaning that cannot be explained by stating the key word, the data. As Einstein once said "learning is experience, everything else is just information" as if he wanted to endorse the empiricist school (Zalta et al., 2003). Machine learning algorithms are binding these two sides in

Table 1: Supervised Learning.

Row	Income level	Age	Experience	Hourly Wage
1	Low	45	11	11.9
2	Low	26	4	14.5
3	High	61	30	37.2
4	Middle	35	9	19.6

Einstein's quote by building a bridge to turn information into experience as emphasized in Mitchell (1997).

Although it is hard to separate into categories, machine learning can be examined in two ways. The first one is clustering the types of machine learning, the second is grouping the main algorithms mostly used.

1.1. Types of Machine Learning

Types of machine learning are classified according to the data whether it has a label or not. So, the data intended to be used will play a great role in selecting a method or defining the problem into one of these methods. According to labeling, there are four types of machine learning problems: supervised, unsupervised, semi-supervised and reinforcement learning.

1.2. Supervised Learning

The simplest and the most easy to understand type of machine learning can be supervised learning. Here, the data used as input is labeled (James et al., 2013). In Table 1, some individuals are labeled according to their income level as can be seen. So, in order to make a prediction from the data by using the variables or features, their labels play a significant role.

A machine learning algorithm should be prepared for the learning process that makes a prediction and corrects these predictions during this learning process. The training keeps iteration until the model achieves a desired level of accuracy on the training, validation or test data.

1.2.1. Regression

Linear regression may be used in linear or non-linear problems where the desired output (label) is a continuous or discrete variable. As the denotation provided here[3], the basic problem to be solved is, given a training set (x, y), where $x = (x_1, ..., x_j)$ is a vector, and represents the (j) input features (independent variables) and (y) represents the output variable (dependent variable). It is intended to predict new $y's$ given new $x's$.

For instance, let's say for some (x), it is being tried to solve for (y). There exists a dataset of (x) values and corresponding (y) results. In order to find the optimal model for this data, it is desired to predict the cost for prediction models and find the model with the least cost.

According to the data and the problem, separation for the solution by using regression is needed:

---

3   Denotations -mostly- as used by Andrew Ng:
   - m: number of training examples
   - x: "input" variables, or features
   - y: "output" variable, or "target" variable
   - $(x^{(i)}, y^{(i)})$: i-th training sample ($i \leq m$)
   - N: number of features
   - X_j : j-th feature ($j \leq n$, but remember to include $x0$)
   - θ: regression parameters
   - H : hypothesis, (a mapping from x to y). E.g. $h_\theta(x) = h(x) = \theta_0 + \theta_1 x + ...$
   - Note that $x_0 = 1$ (So that it is obtained as $h_\theta(x) = \vec{\theta}^T x = \theta_0 x_0 + \theta_1 x_1 + ...$)
   - J: cost function to minimize. Regression is to minimize $J(\vec{\theta}) = \frac{1}{2m} \Sigma_{\forall i} \left(h_\theta(x^{(i)}) - y^{(i)} \right)^2$

Linear Regression: If regressand/response/dependent variable or output feature shows a linear relationship between independent/explanatory variables or features, then the best solution for this problem is linear regression. In here, sum of squared errors (SSE) function $(y - XB)^2$ should be a convex function that can be represented as linear. If not, the solution for this kind of problem is different from here.

Non-linear Regression: SSE function may show non-convex properties. That is, the continuous response variable shows curvilinear relationship between at least one independent variable. These problems may contain multiple local minima. Due to nonlinearity between variables/features, the local minima can be achieved by using numerical optimization algorithm to determine coefficients. The most common non-linear regression forms are power, weibull and fourier equations.

1.2.2. Classification and Logistic Regression

Another problem that needs to be solved is the classification problem in which the data has a labeled/output variable that are not continuous, but a discrete, more specifically binary/dichotomous. In such cases, linear methods will perform standard model prediction. This model is actually the possibility of the realization of the dichotomous variable. However, in linear models, this probability increases or decreases linearly with the dependent variable. Even the total probability may exceed 1. Since it is impossible to estimate, specifically the probability of occurrence of something by using linear methods, the most appropriate method for them is logistic classifiers. This is so similar in econometrics the discipline used to estimate the probability of occurence as a discrete choice by using generalized linear methods family. In short, logistic classifiers are;

- used to solve classification problems (desired output: classifier variable)
- the most proper methods since linear regression is not advised for classification. First, it is sensitive to outliers/skewed data sets, second, probability is not linear and cannot exceed 1.

Instance-based Learning: Machine learning models which are instance-based generally create similarity (or dissimilarity) measurements to calculate representative power of new instances when they append to the data. This type of learning is also called memory-based learning since it redraws the hypothesis to make a new distance or similarity index by recalling old instances. It is mostly used in recommendation systems. Certain types of instance-based machine learning algorithms are:

– k-Nearest Neighbor (kNN),
– Learning Vector Quantization (LVQ),
– Self-Organizing Map (SOM),
– Locally Weighted Learning (LWL),
– Radial Basis Function Network (RBFN).

1.3. Unsupervised Learning

In this type of machine learning problem, the data has no output or label (Friedman et al., 2001). The first aim is to regenerate the data structure or redraw the distribution that is targeted. Since there are no correct labels or outputs to estimate the accuracy of the machine learning model, it is unrealistically hard to make sure of its prediction power. In Table 1, suppose that the labels in the feature "Income level" are unknown, this turns it into the problem that should be solved with unsupervised learning. The first thing to do in order to generate income level groups, is to use other features to remap the clusters, but first the association between other features could be examined to make clustering easy.

1.3.1. Clustering

It is said that this is the first unsupervised learning algorithm. In unsupervised learning problems, there only exists an input data $x^{(i)}$ with no labels, and it is intended for the algorithm to find some structure in the data. A clustering algorithm such as the k-means algorithm attempts to group the data into k "clusters" which have some similarity. Some examples include: market segmentation, social network analysis, organizing computer clusters, and astronomical data analysis.

First, it could be done to choose how many clusters are wanted, k [4]. This could be done by "eyeballing" the data, or from *a priori* hypotheses about how many clusters there should be. k-means clustering algorithm consists of two parts. First, it is initialized with the cluster centroids $\mu_1,...,\mu_k \in R^n$ randomly. One way is to set the cluster centroids to be equal to a random subset of the training examples. Note that k-means can get stuck in local optima, so one solution is to try multiple times with different random initializations.

---

4   It should go without much saying that the number of clusters k should be less than the number of training examples m.

1.4. Semi-supervised Learning

Semi-supervised data, as can be understood from its name, cover a mix of labeled and unlabeled data (Friedman et al., 2001; Haykin, 2009). In general, these mixed data for semi-supervised machine learning consist of mostly unlabeled ones when comparing its proportion to the amount of labeled data. It may be done in the nature of the data, which can be dealt with in the semi-supervised learning. But somehow, labels of the data might be collected after other features which have been completed. In order to save time, with decreasing the cost of the data of labels which are missing, semi-supervised learning will do it with a remarkable success.

1.5. Reinforcement Learning

In economic terms, reinforcement learning needs a data shaped among the environment, players, and moves which can be rewards or punishments (Haykin, 2009) according to the set of strategies, all of the concepts recall the game theory. The setting of reinforcement learning is not the only part similar to the game theory, but the goal is also quite similar to the game theory, the actions played must get maximized rewards at the end of the play, that is, the data generating process.

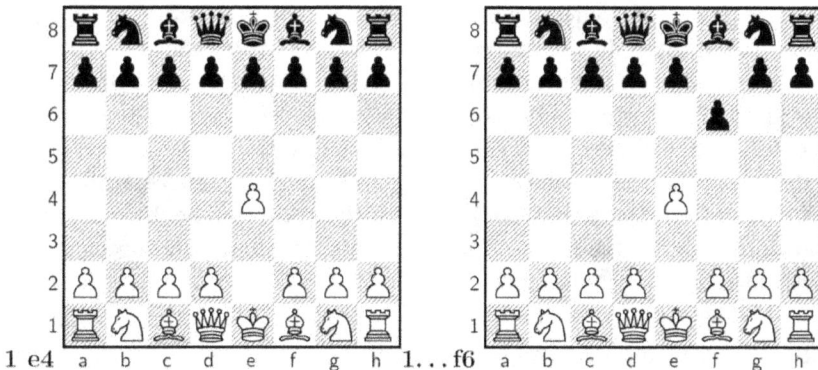

Figure 1: Reinforcement Learning

The data being created dynamically can be exemplified by a chess game. As drawn a chess in Figure 1, the white player is the first mover. After that, black player plays f6 against white's e4 move. Cumulative rewards can be maximized by minimizing the pieces lost to other player or maximizing the pieces kept to win the game. If the pieces have no equal points affecting the rewards, second problem should be turned into minimization of cumulative points of pieces. Ng et al. (2006) design a good and innovative implementation of reinforcement learning as an example.

1.6. Federated Learning

Learning methods studied previously are closely related to the type of data. Federated learning, however, relies on the shape of data. As McMahan et al. (2016) proposed, the data can be stored into small pieces and decentralized, however, the learning updates can be done by local computers so that make the machine learning model is as correct as possible. The first attempts come from Google which collects the data from smartphones to make smarter than its initial settings by using Android OS and trains the models at the center, after a while, it sends the trained model by updating initial settings of the OS.

1.7. Transfer Learning

Like federated learning, transfer learning is a newcomer in the field of machine learning. A model can be trained by using data which have certain attributes (features, variables). When the model reaches a desired accuracy, it is accepted as learned weights and adjusted parameters as the learning process is finalized. A new data which is not the continuation of the old one, however, the new one provides similar settings, collection methods and especially attributes compared to older one. Finally, the learning algorithm can be transferred into the new data with small adjustments which shorten the period rather than starting a new learning process. Dai et al. (2007) assert labeling tasks can be the most convenient jobs for being a subject of transfer learning that boosts learning and are less costly than unsupervised learning.

1.8. Ensemble Learning

Ensemble learning is mostly suitable for the problems classified in supervised machine learning, where there must be at least two or more complex machine learning algorithms to estimate one single hypothesis. However, ensemble learning can give multiple outputs which may be mixed to increase accuracy of prediction. In machine learning literature, there are two main common ensemble methods, bagging and boosting (Dietterich, 2000). The last follower is stacking.

Bagging: The first method for ensemble is bagging which stands for bootstrap aggregation. It is a way to reduce the variance of an estimate by averaging all estimates. Bagging, at first generates bootstrap sampling to create subsets for training base learners. Then, bagging votes for classification and weighting for regression.

Boosting: The second method is called boosting. Boosting transforms weak learning algorithms to stronger ones. Sequentially estimated weak learners are weighted. Then, a majority vote classification or weighted sum rule are run to produce the final estimate.

Stacking: The other method is stacking which is abbreviated from stacked generalization. There is a meta learner that learns the output of a sample from the outputs of the base learners in stacking. The base models are trained according to whole training set. Later, meta learner is trained on the outputs of base model as independent variables (features).

2. Neural Networks

Neural networks are one of the most popular methods used in research and applications according to Jain et al. (1996). It is used in many applications such as optical character recognition (the automation of the postal service) and credit card authentication. It allows an easy way to compute non-linear classification problems, because essentially at the heart of the algorithm, one lets it compute its own parameters and weights[5].

Neural networks are inspired by biology, notably by brain plasticity, which is the ability of different parts of the brain to learn new functions, such as a classic neural re-wiring experiment where the auditory/somatosensory cortex can learn to "see" (receive and process visual input). The brain's amazing ability to adapt different regions of the brain to learn the functions of another part suggests a "one learning algorithm" hypothesis, rather than sense-specific algorithms. In this sense, a neuron can be modeled as a computing unit, with dendrites as "input" wires and the axon as an "output" wire.

In the neural network model, a neuron is a single unit with an activation function (i.e. sigmoid/logistic). A network consists of different layers of neurons. The first layer, which directly receives the input, is called the **input layer**. The first layer then feeds into the second layer, etc., until the final layer, which eventually computes the estimation, is called the output layer. The rest of the layers in between are called the hidden layers.

Let it define $a_i^{(j)}$ as the *activation* of unit i in layer j, i.e. $a_i^{(j)} = 1$ if unit i in layer j fires. (Similar to logistic and linear regression, it can be added a constant term, a_0, also called a **bias** term). Let $\theta^{(j)}$ be the matrix of weights that control the mapping from layer j to layer $j+1$. If the network has s_j units in layer j and s_{j+1} units in layer $j+1$, then $\theta^{(j)}$ will be of dimension $s_{j+1} \times \left(s_j + 1\right)$ because of the bias term.

---

5   The problem is that most of the time, it's a non-convex optimization problem (so gradient descent is not guaranteed to work). Also, it is computationally expensive to train a neural network. SVMs are more reliable. That being said, SVMs have their own pros and cons too.

Passing information through the neural network from input layer to output layer, called **forward propagation**, is achieved via,

$$a_k^{(2)} = g\left(\theta_{k,0}^{(1)} x_0 + \theta_{k,1}^{(1)} x_1 + ...\right) \tag{1}$$

$$= g\left(\theta_{k,0}^{(1)} a_0^{(1)} + \theta_{k,1}^{(1)} a_1^{(1)} + ...\right) \tag{2}$$

$$= g\left(\theta^{(1)} a^{(1)}\right) \tag{3}$$

or more generally,

$$= g\left(z^{(j+1)}\right); \ z^{(j+1)} = \left(\theta^{(j)} a^{(j)}\right) \tag{4}$$

Note that, similar to logistic regression, neural networks can similarly be adapted to multi-class classification. One way is similar to the one-vs-all algorithm: but now, it can be specified that output to be a vector. For example, you can write your algorithm such that the output $h_\theta(x) = \begin{bmatrix} 1000... \end{bmatrix}$ if first class, $\begin{bmatrix} 0100... \end{bmatrix}$ if second class etc., and that training set would just be $\left(x^{(i)}, y^{(i)}\right)$ where $y^{(i)}$ is $\begin{bmatrix} 1000... \end{bmatrix}$ or $\begin{bmatrix} 0100... \end{bmatrix}$. etc. depending on class.

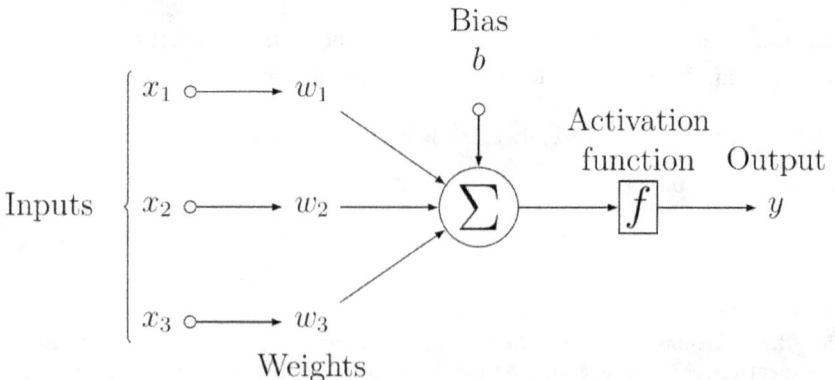

Figure 2: Single Layer Artificial Neural Network (ANN).

2.1. Single Layer Perceptron

If an ANN structure has one layer that contains neurons fed by weighted inputs, it is called a single layer artificial neurons. A simple architecture of single layered neuron is shown in Figure 2. As stated in the comparison of machine learning and econometrics below, any econometric problem can be thought of as a one layered machine learning problem without considering its iterative process.

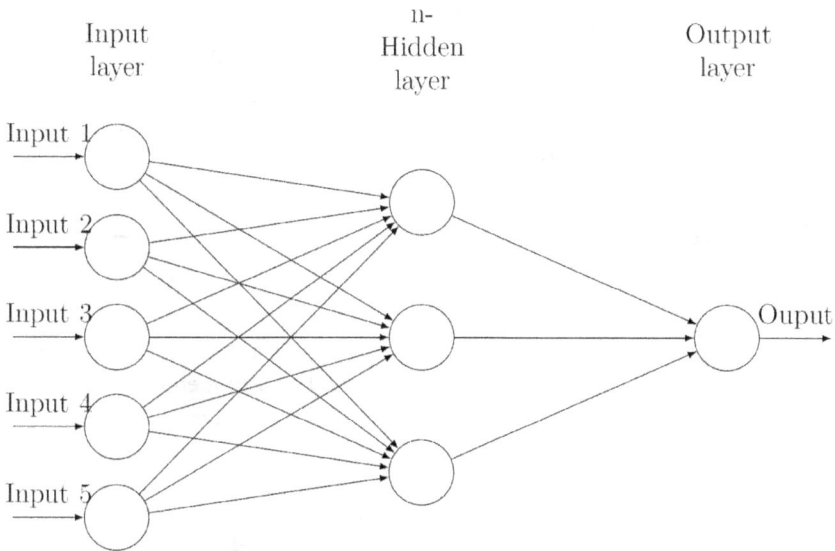

Input layer   n-Hidden layer   Output layer

Input 1   Input 2   Input 3   Input 4   Input 5   Ouput

Figure 3: Multi-layer (Deep) Artificial Neural Network.

2.2. Deep Neural Networks

Deep Neural Network is similar to single layer neural networks except for its hidden layer. If the number of hidden layers is at least 1, any shallow or wide machine learning architecture turns out to be deep learning (DL) architecture as drawn in Figure 3.

In DL, there are no feedback connections in which outputs of the model are fed back into itself. The model is associated with a directed acyclic graph describing how the functions are composed together. These networks require initializing all weights to small random values. To apply gradient-based learning, a cost function and output representation must be chosen. Most modern NNs are trained using maximum likelihood. This means that the cost function is simply the negative log-likelihood:

$$J(\theta) = -\mathbb{E}_{x,y \sim \hat{y}data} \log p_{model}(y \mid x) \tag{5}$$

Most hidden units can be described as accepting a vector of inputs (x), computing an affine transformation $(z = W^T x + b)$ and then applying an element-wise nonlinear function $(g(z))$.

Most NNs are organized into groups of units called layers. Most NN architectures arrange these layers in a chain structure, with each layer being a function of the layer that preceded it. In these architectures the main considerations are choosing the depth of the network and the width of each layer. Deeper networks generalize better (most of the time).

2.3. Architecture Design

First, it would be needed to choose a particular network architecture (how many layers, how many units...). Usually, the dimension of the input (data) and output (e.g. k classes) are decided by problem. A reasonable default is having 1 hidden layer. If one chooses to implement more than 1 hidden layer, it's a good rule of thumb to have the same number of hidden units in every layer. Usually, the more units you have in each layer, the better, but it gets more expensive to compute.

Start for the ANN to run should be as follows:

– Random Initialization of weights,
– Forward Propagation,
– Compute Cost,
– Back Propagation (check with gradient checking, then turn off),
– Optimize (e.g. using gradient descent) cost function.

2.3.1. Feed Forwards

An FNN[6] is formed by one input layer, one (shallow network) or more (deep network, hence the name deep learning) hidden layers and one output layer. Each layer of the network (except the output one) is connected to a following layer. This connectivity is central to the FNN structure and has two main features in

---

6   FNN stands for Feedforward Neural Networks.

its simplest form: a weight averaging feature and an activation feature. It will be reviewed these features extensively.

2.3.2. Convolutional Neural Networks

Convolutional Neural Networks (abbreviated as CNN in ML literature) implement the idea of signal convolution to neural networks. It actually separates image data into the fragments in many ways on neurons, filters and then applies locally on subregions of the images rather than of the whole image (Lecun et al., 1998). The filters derived from these pre-processes are convolved on input images, then, it is used for creating new features, feature maps, intermediate maps.

2.3.3. Sequence Modeling

In general, sequence modeling differs from other types of machine learning models since the output estimated in a neural network feed other perceptron sequentially according to the architecture of the system. As a result, sequence modeling is the most appropriate model for time series problems in econometrics.

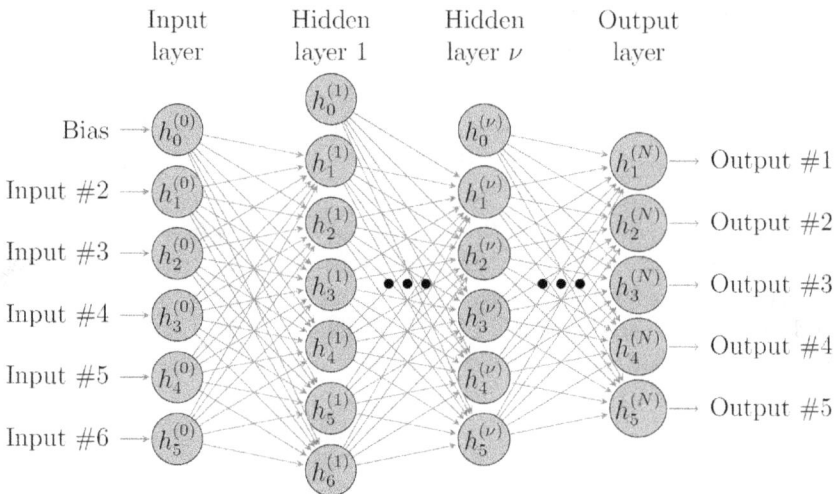

Figure 4: Neural Network with $N+1$ Layers. Notations are simplified, the index referencing the training set has not been indicated. Shallow architectures use only one hidden layer. Deep learning amounts to take several hidden layers, usually containing the same number of hidden neurons. This number should be on the ballpark of the average of the number of input and output variables.

2.3.3.1. Recurrent Neural Networks

Recurrent neural networks (RNN) are networks in which connections between units form a directed loop. It is allowed to exhibit dynamic temporal behavior with RNN. Unlike feedforward neural networks, RNNs can use their input memory to process inputs. This attribute makes RNNs a useful method of handwriting recognition, speech recognition and time series forecasting.

People don't learn new meanings for a word they've just learned. They relate to the word they have just learned by starting from similar pre-existing words. However, traditional artificial neural networks do not have this meaning attribute found in humans, and this is their greatest deficiency. For example, if you want to classify activities by looking at all the frames in the video, traditional neural networks will not be able to classify because they cannot make sense between people like frames.

RNNs, by creating a loop, will provide the use of historical information and thus can make classification by making meaning between frames. In traditional artificial neural networks, the results from the cells do not come back to them as input. In RNN, the result from the cell comes back to it as input. If RNN is opened, an architecture as shown in the following figure appears. In the time frame, the same cell repeats itself more than once. Thus, meaning can be established between frames.

Forward Pass in an RNN-LSTM: In Figure 5, the RNN architecture is presented in a schematic way; the time dimension is of size 8 while the "spatial" one is of size 4. The real novelty of this type of neural network is the fact that it is being used to predict a time series encoded in the very architecture of the network. RNN have first been introduced mostly to predict the next words in a sentence (classification task), hence the notion of ordering in time of the prediction. But this kind of network architecture can also be applied to regression problems. Among other things one can think of are stock price evolution or temperature forecasting.

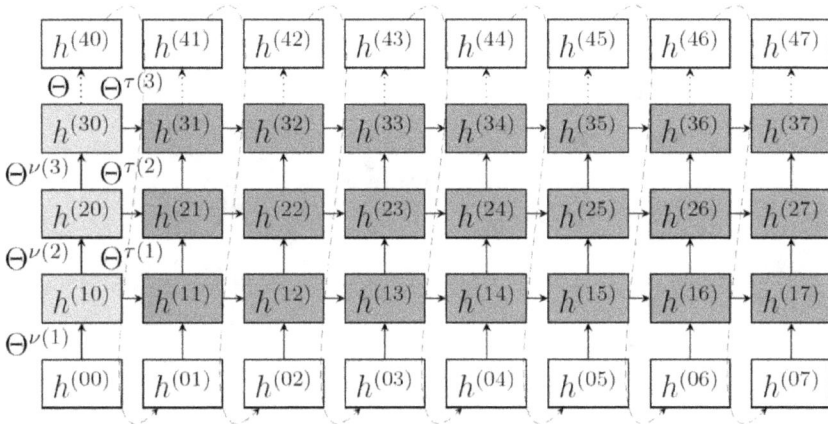

Figure 5: RNN Architecture within Space and Time.

RNN and LSTM / GRU: All RNNs have feedback loops in the recurrent layer. This lets them maintain information in "memory" over time. But, it can be difficult to train standard RNNs to solve problems that require learning long-term temporal dependencies. This is because the gradient of the loss function decays exponentially with time (called the vanishing gradient problem). Long Short-Term Memory (LSTM) networks are a type of RNN that uses special units in addition to standard units. LSTM units include a "memory cell" that can maintain information in memory for long periods of time. A set of gates is used to control when information enters the memory, when it's output, and when it's forgotten. This architecture lets them learn longer-term dependencies. Gated recurrent units (GRUs) are similar to LSTMs but use a simplified structure. They also use a set of gates to control the flow of information, but they don't use separated memory cells, and they use fewer gates.

In an LSTM Neural Network, the state of a given unit is not directly determined by its left and bottom neighbors. Instead, a cell state is updated for each hidden unit, and the output of this unit is a probe of the cell state. This formulation might seem puzzling at first but it is philosophically similar to the ResNet approach. Instead of trying to fit an input with a complicated function, it is tried to fit tiny variation of the input, hence allowing the gradient to flow in a smoother manner in the network.

3. Activation Functions

Artificial neural networks are indispensable to the activation functions. There are three main categories in ANN literature: Monotonic activation functions which are the most common functions used in ANN is the first one. Periodic activation functions are the second group of activation functions, they are less known functions which are hard to compute to meet special needs of the architecture of ANN, but may perform well especially in Adaptive Artificial Networks. The last part of activation functions is actually neither an activator nor a function, nevertheless, it is an essential part of activation function, it is called bias unit that is the last chapter.

3. Activation Functions

4. Monotonic Activation Functions

In mathematics, monotonicity of any arbitrary function describes its nonde-creasing or nonincreasing feature (Royden and Fitzpatrick, 1988). Visually, in Figure 6, thick with dotted function gradually decreases (nonincreasing) while thick function mostly increases (nondecreasing).

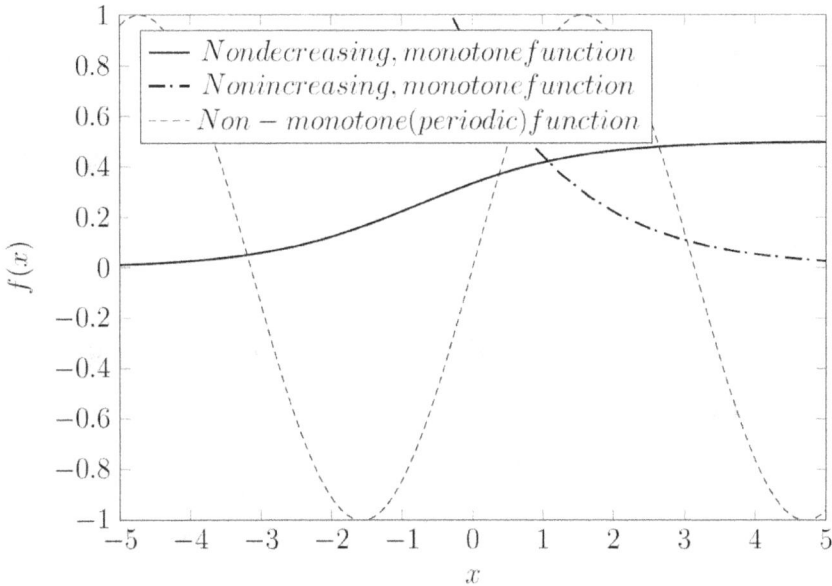

Figure 6: Monotonic Functions.

Conceptually, if a function satisfies that if $\forall x, y \in \mathbb{R} \, and \, x \geq y$, then $f(x) \geq f(y)$ is valid for a certain bound without changing the order, which implies nondecreasingness of the function f. Conversely, $\forall x, y \in \mathbb{R} \, and \, x \leq y$, so the relationship such as $f(x) \leq f(y)$ is preserved for previously determined bound without changing the order, that is the functions which are nonincreasing are at least weakly monotonic.

4.1. Linear Function

If an input is thought to be shaped according to weights determined either endogenously or exogenously (that feed neurons as proportionally determined by hand) to neuron, linear function may be the first considered.

The main drawbacks for this function are two with high priority: First one is its derivative of it which is 1 and the results in that gradient descent are full of 1s and make it a constant gradient that indicates there is no actual relationship with x and gradient descent. Second one, the errors coming from prediction should be corrected with a constant by backpropagation showing there is no dependency in the changes in input (that is $\delta(x)$). The general formula can be stated as following with a small restriction:

$$f(x) = \alpha x, \tag{6}$$

Where $\alpha \neq 1$, in the range of $(-\infty, \infty)$, continuous in C^∞, monotonic for itself and first derivative.

If the first parameter, $\alpha = 1$, it simply turns out to be an identity function which is explained in later. Both of these situations are drawn in the Figure 7.

4.1.1. Identity Function

If $\alpha = 1$, it becomes an identity function. At first sight, as stated in Section 4.1, an identity function may seem useless since it does not perform any transformation of inputs to neuron. But, actually, it transforms itself because the neuron uses it with no weights. So, an identity function can be called as replicator or duplicator for summation of weights ($\Sigma_j w_{kj} x_j$, namely summing function (Haykin, 2001)) with no weights (Rice, 1953). The general formula given below is also represented in Figure 7.

4.1.2. Piecewise Linear Function

Piecewise linear function in Equation (7) restricts the inputs between α_{min} and α_{max} in order to produce the values in between 0 and 1 as outputs predicted (Zeng et al., 2010) by dropping other values of inputs where it gives 0 if it is lower than minimum threshold or produces 1 if it is higher than maximum as drawn in Figure 8.

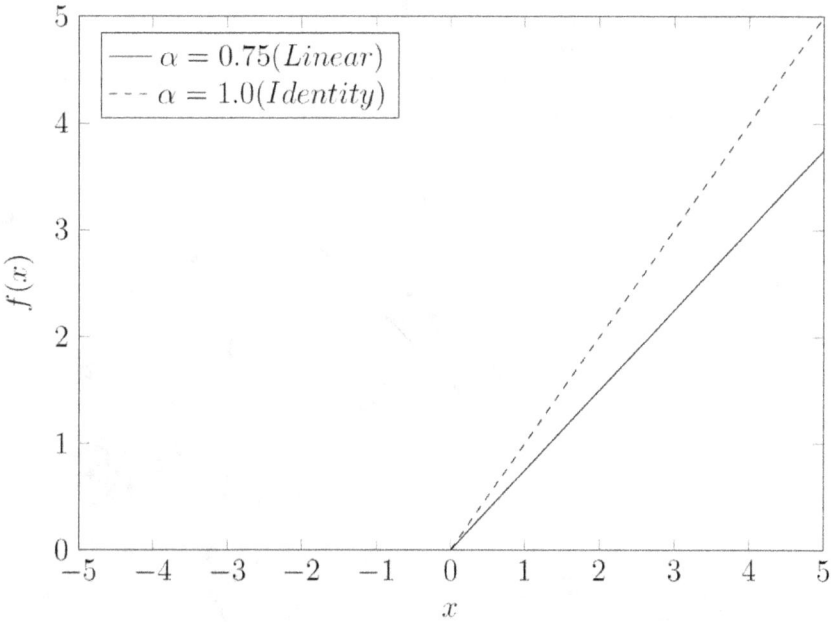

Figure 7: Linear and Identity Functions.

$$f(x) = \begin{cases} 0 & \text{If } x < a_{min} \\ mx + b & \text{If } a_{min} \leq x \leq a_{max} \\ 1 & \text{If } x > a_{max} \end{cases} \tag{7}$$

where,

$$m = \frac{1}{a_{max} - a_{min}} \tag{8}$$

and,

$$b = -ma_{min} = 1 - ma_{max} \tag{9}$$

in the range of $(-\infty, \infty)$, continuous in C^∞, monotonic for itself and first derivative.

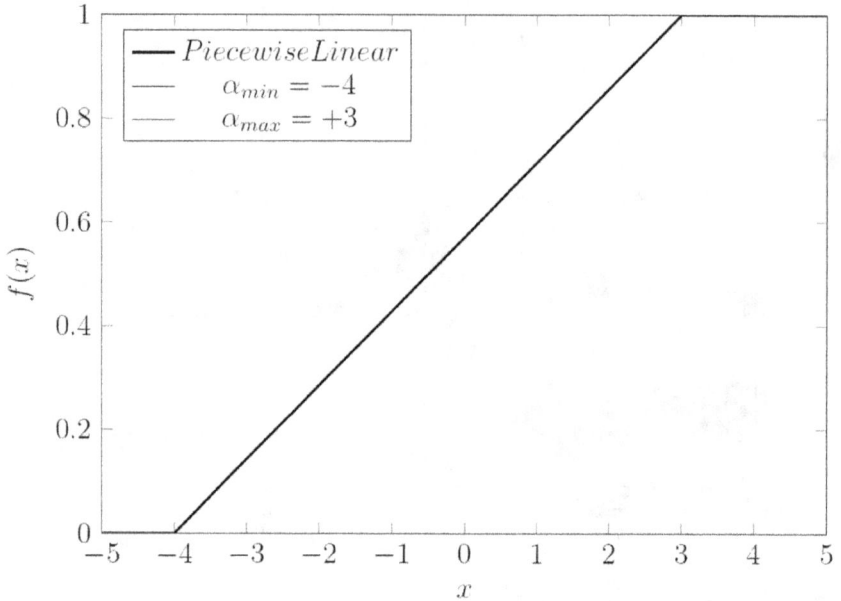

Figure 8: Piecewise Linear Function.

4.2. Threshold (Unit Heaviside, Binary, Step) Function

Well known threshold function (Batres-Estrada, 2015) has three other synonyms as unit Heaviside (Osher and Fedkiw, 2003; Cox, 1992), binary and step function (10). In econometrics literature, it is better known as names, it is similar to a dummy variable which can be used alone or interactively with weights as its estimated coefficient. The usefulness of a threshold function comes from its filtering power of inputs and feeding neuron like a bias term; econometrically, initially it helps to change constant term, consequently, it changes $\hat{y}$ predicted. Following is the threshold function and is represented in Figure 10:

$$f(x) = \begin{cases} 1, & if : x \geq 0, \\ 0, & if : x < 0. \end{cases}$$

in the range of $(0, 1)$, continuous in C^{-1}, monotonic for itself.

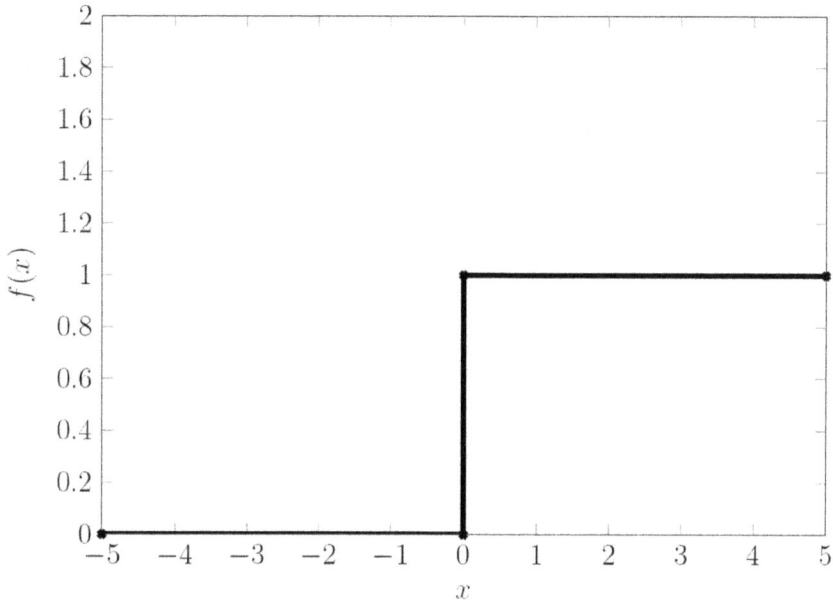

Figure 9: Threshold (Unit Heaviside / Binary / Step) Function.

4.3. Sigmoid Function

The sigmoid function is one of the most (Friedman et al., 2001; Batres-Estrada, 2015) common types of activation functions used in neural networks, especially in solving classification problems. It is generally used for linear and nonlinear transformation of inputs in order to feed other neurons as balancer. The general formula stated below is represented in Figure 10:

$$f(x) = \frac{1}{1+e^{-ax}},\tag{11}$$

in the range of $(0,1)$, continuous in C^{∞}, monotonic for itself.

The sigmoid function (11) was first invented by Verhulst (1838) and used for estimating population growth (Verhulst, 1977)[7] as a logistic function or logistic curve. In econometric terms, if the α parameter[8] equals one, it is called a logistic function (Greene, 2003) and used for calculating probabilities when the dependent variable is a dichotomous one. Since it can also be classified in antisymmetric functions comparing origins, backpropagation process may converge faster in a training process (Haykin, 2001; LeCun et al., 2012).

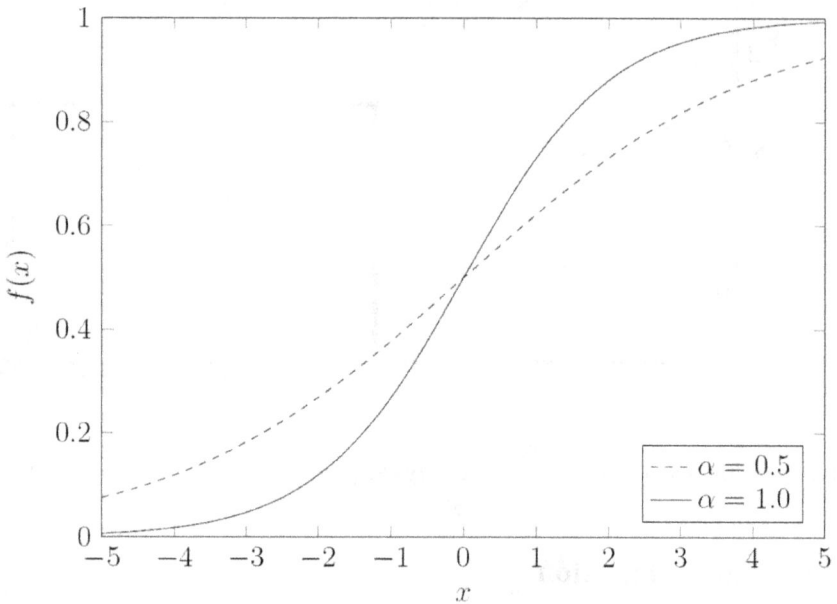

Figure 10: Sigmoid Function.

4.3.1. Bipolar Sigmoid Function

Bipolar sigmoid function is a sigmoid function with a small difference; it produces values in the range of (-1,+1). So it cannot be used for probabilistic estimations since probability cannot be lower than zero. Panicker and Babu (2012) find that

---

7   Reprinted edition.
8   It is called steepness level of sigmoidals.

a bipolar sigmoid is more efficient than other sigmoidals. The general formula is shown in Figure 11 and is given below Equation 12:

$$f(x) = \frac{1 - e^{-ax}}{1 + e^{-ax}}, \tag{12}$$

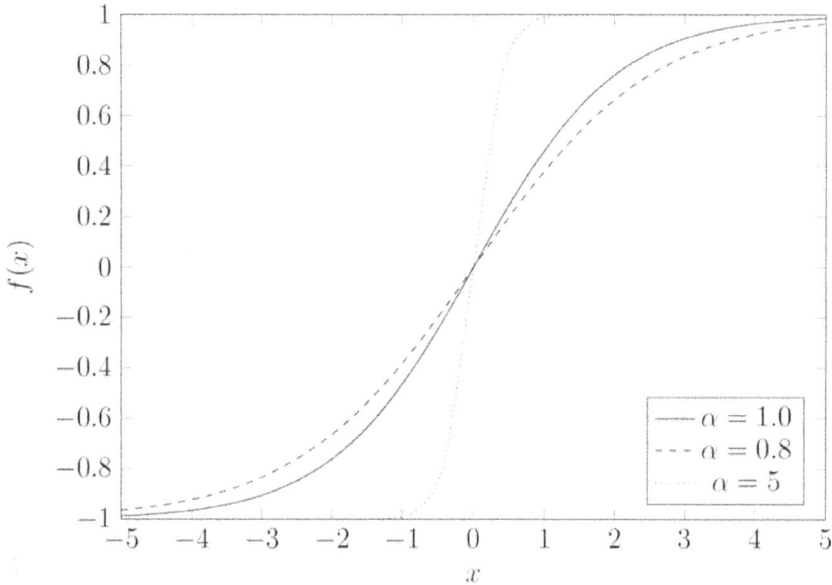

Figure 11: Bipolar Sigmoid Function.

where α is the smoothness parameter of a bipolar sigmoid that an activation function can be defined in the range of $(-1, +1)$, continuous in C^∞, monotonic for itself.

4.4. Rectified Linear Unit (ReLU)

Via the abbreviated and capitalized name, ReLU is a recent newcomer called in the existing literature as Ramp function and advanced function (Nair and Hinton, 2010) that provides numerous replications from the sigmoidal function (Batres-Estrada, 2015) with the same learned weights and biases determined in the training process. The formula is given blow as shown in Figure 10:

$$f(x) = \begin{cases} ax, & if : x < 0, \\ x, & if : x \geq 0. \end{cases} \tag{13}$$

where $a = 0$, in the range of $[0, +\infty)$, continuous in C^0, monotonic for itself and its derivative.

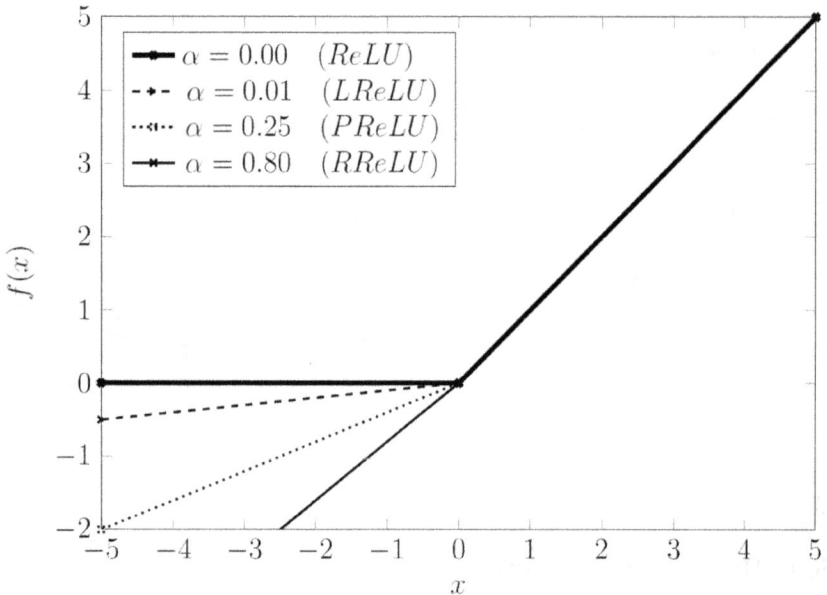

Figure 12: Rectified Linear Unit (ReLU) Family.

As stated in Equation (13), representation of RELU does not cover an α parameter, this is put there to easily identify other family members of RELU. Figure 12 displays only the approximation of RELU, as it was stated previously, it replicates a large number of sigmoids. This is why it is different from a linear function, the nature of RELU is nonlinear. If the neural network is so complex, having tens or hundreds of neurons, most of them may be affected by other activation functions such as tanhs, softpluses, sinusoids, sigmoids, and Gaussians.

RELU is invented to burst the speed of training (Nair and Hinton, 2010) and make it easier to generalize properties of most of the linear models (Goodfellow et al., 2016).

4.4.1. Leaky Rectified Linear Unit (LReLU)

The first member of the ReLU family of functions is the Leaky Rectified Linear Unit (LReLU) which is simply defined as Equation (13) when the parameter $\alpha = 0.01$, in the range of $(-\infty, +\infty)$, continuous in C^0, monotonic for itself and its derivative (Maas et al., 2013). It is shown in Figure 12.

4.4.2. Parametric Rectified Linear Unit (PReLU)

The second member of ReLU is the Parametric Rectified Linear Unit (PReLU) which can be represented as Equation (13) without a restriction on α parameter, where in the range of $(-\infty, +\infty)$, continuous in C^0, monotonic for itself, if and only if $\alpha \geq 0$, and its derivative. It is shown in Figure 12.

4.4.3. Randomized Rectified Linear Unit (RReLU)

As the name of this member of the ReLU family explains its function, Randomized Rectified Linear Unit (RReLU) randomly selects an α parameter from $(-\infty, +\infty)$ as declared in Equation (13) without any restriction (Xu et al., 2015). The main flaw of RReLU is that the difference between PReLU and RReLU cannot be identified when a backpropagation process is in input (that is $\delta(x)$). It is defined where in the range of $(-\infty, +\infty)$, continuous in C^0, monotonic for itself, if and only if $\alpha \geq 0$, and its derivative. It is drawn in Figure 12.

4.5. Exponential Linear Unit (ELU)

Exponential Linear Unit (ELU) is another variant of ReLU and its family but the second statement presented in Equation (14) is totally disparate from ReLU and its family (Nair and Hinton, 2010). It is proven that ELU is one step ahead of ReLUs explained previously (Clevert et al., 2015; Trottier et al., 2016). It converges the mean of activation functions close to zero, the process decreases time required to train a dataset containing particularly computer visions tasks.

$$f(x) = \begin{cases} a(e^x - 1), & if : x < 0, \\ x, & if : x \geq 0. \end{cases} \tag{14}$$

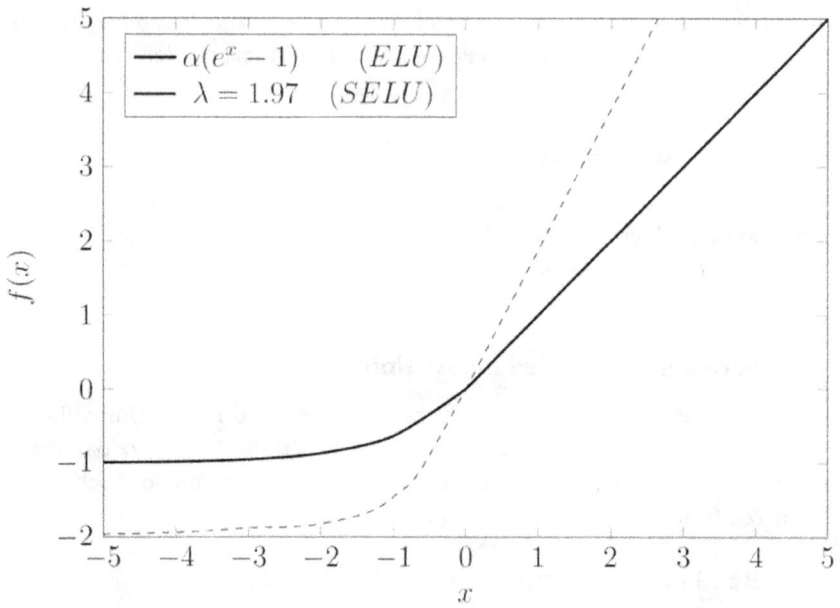

Figure 13: Exponential Linear Unit (ELU).

In the range of $(-\alpha,+\infty)$, continuous in C^1 if $\alpha=1$, otherwise in C^0, monotonic for itself, if and only if $\alpha \geq 0$, and its derivative iff $\alpha \in [0,1]$. As shown in Figure 13 and stated above in Equation (14), ELU's limit of converging $-\infty$ will be -1. That is, negative values of x doesn't affect it.

4.5.1. Scaled Exponential Linear Unit (SELU)

Scaled Exponential Linear Unit (SELU) is derived from ELU with a λ that may be determined exogenously by hand or endogenously by the speed of self-normalization of neuron (Klambauer et al., 2017). It permits to run recursively normalization of a neuron itself and to deal with the vanishing gradient descent (Clevert et al., 2015). SELU is shown in Figure (14) and is stated below as Equation (15):

$$f(x) = \lambda \begin{cases} a(e^x - 1), & if : x < 0, \\ x, & if : x \geq 0. \end{cases} \tag{15}$$

4.6. SoftMax Function

As previously explained, a dichotomous dependent variable that needs to calculate probabilities should be estimated with sigmoid/logistic (in the Equation 11) functions. If there are multiple, but discrete choices, in a dependent/output variable, it is basically estimated with using a multinomial logistic function which is frankly a generalization of sigmoid functions (Friedman et al., 2001). In machine learning literature, most of the sigmoidals are generalized into SoftMax function in order to calculate multiple outcomes for each choice as a vector of probabilities. Well known function in machine learning literature is given below:

$$f_i(\vec{x}) = \frac{e^{x_i}}{\sum_{i=1}^{J} e^{x_j}} \tag{16}$$

in the range of $(0, +1)$, continuous in C^{∞}.

SoftMax function is not only popular in artificial neural networks, but also frequently used in Naïve Bayes, linear discriminant analysis and multinomial logistic regression. The only, and tiny difference, statistically it generates a significant differentiation, is that these multiple discrete choices must be mutually exclusive classes, there should not be any dependency between the classes processed with the SoftMax function.

4.7. Odd Activation (Signum, Sign) Function

As formulated below, the odd activation function is as follows Equation 17 drawn in Figure 14:

$$f(x) = \begin{cases} -1, & if \ x < 0, \\ 0, & if \ x = 0, \\ 1, & if \ x > 0. \end{cases} \tag{17}$$

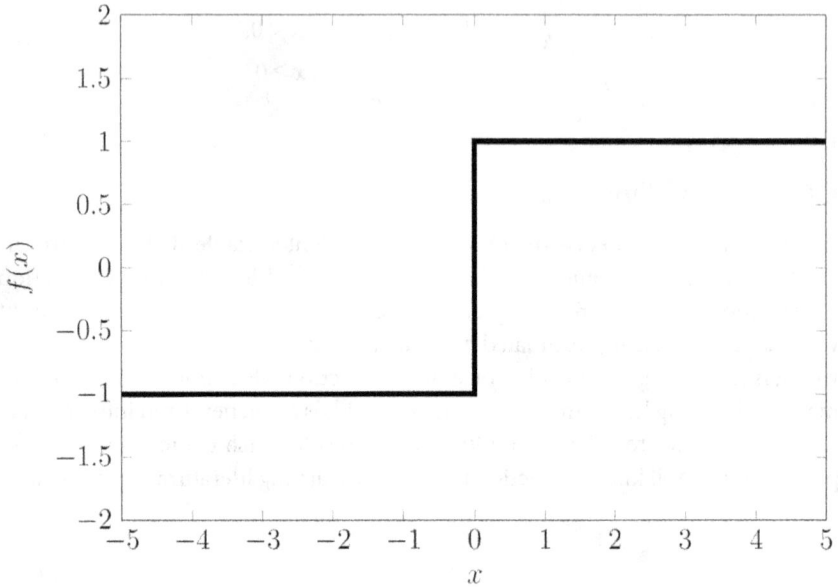

Figure 14: Odd Activation (Signum / Sign) Function.

where $\alpha = 0$, in the range of $(-\infty, +\infty)$, continuous in C^0, monotonic for itself and its derivative.

4.8. Maxout Function

Hinton et al. (2012) are looking for a way to reduce training time of models consisting of a few ensemble models and find the technique called Dropout while it is also examined that estimated predictions should be harmonized by averaging them. Goodfellow et al. (2013) offer a new feed-forward artificial neural network model including a recently invented activation function called maxout where the activation function is the maximum of the inputs fed. It is so useful with Dropout (Goodfellow et al., 2013). The form of this activation function is given below:

$$f(\vec{x}) = max_i x_i \tag{18}$$

where in the range of $(-\infty, +\infty)$, continuous in C^0.

Goodfellow et al. (2013) provide that when the number of epochs exceeds approximately 30, maxout maintains significant improvements in the large cross validation set. Also, model averaging on MNIST classification using maxout function beats other rectifiers.

4.9. Softsign Function

Some functions are useful when they play role by balancing two or more functions and may be better tools. The softsign, which is also sigmoidal, is such a function and looks like a smoothed version of both tanh in Equation 21 (with high weight) and sigmoid function, and a bit odd activation function (with less weight). It also approximates identity function with its derivate at origin (Aghdam and Heravi, 2017). In some instances, it may be considered an alternative to tanh function as it does not saturate as easily as hard clipped functions. The function is stated in Equation (19) below and visualized in Figure 15:

$$f(x) = \frac{x}{1+|x|}, \tag{19}$$

in the range of $[-1,+1]$, continuous in C^∞, monotonic for itself.

The most significant property of the softsign function is the saturation level which will be higher as $|x|$ increases, saturation level in comparison with tanh is a bit softer compared to other functions. This property makes it compute at ease.

4.10. Elliott Function

Elliott function (Elliott, 1993) looks like another sigmoidal which produces results between 0 and +1 for output processed with Elliott. It performs very-well and does not suffer from neuron vanishing problems according to applications done in Matlab (Ploskas and Samaras, 2016). Its formula is given below Equation (20) and drawn in the Figure 15.

$$f(x) = \frac{0.5(x)}{1+|x|} + 0.5 \tag{20}$$

in the range of $[0,+1]$, continuous in C^∞, monotonic for itself.

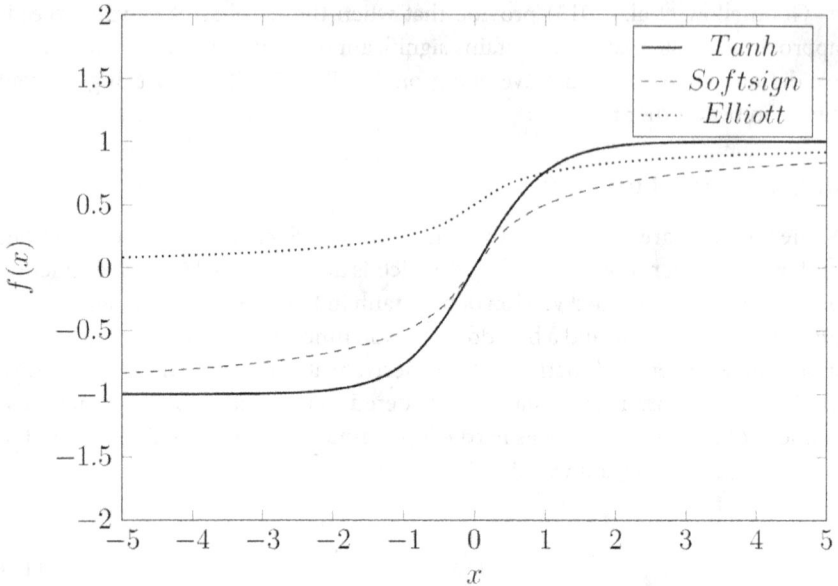

Figure 15: Softsign, Hyperbolic Tangent and Elliot Function.

4.11. Hyperbolic Tangent (Tanh) Function

Sauri (1774) first introduces the hyperbolic tangent function, which is another sigmoidal, in short "tanh". It becomes another well-known function in machine learning literature. Although it comes from $\dfrac{sinh(x)}{cosh(x)}$, the shape of it resembles sigmoid and softsign functions a bit, it actually is a rescaling of a sigmoid function. Tanh and its rescaling procedures are given in Equation (21, 22, 23) and are illustrated in Figure 15:

$$f(x) = \tanh(x) = \frac{2}{1+e^{-2x}} - 1, \tag{21}$$

where in the range of $(-1, +1)$, continuous in C^{∞}, monotonic for itself. Which can also be derived from a sigmoid function;

$$\tanh(x) = 2.\sigma(2x) - 1, \tag{22}$$

where $\sigma(x)$ is,

$$\sigma(x) = \frac{e^x}{1 + e^x}, \tag{23}$$

The most valuable contribution of tanh and sigmoid for machine learning literature is that they can map any real number determined in the range of $(-\infty, +\infty)$ to a number in between $[-1, +1]$ with tanh or $[0, 1]$ with sigmoid function respectively. These procedures are called normalization which makes internal covariates accurate predictors by generating batch normalization so as to normalize each vector having zero mean and unit variance. LeCun et al. (2012) also point out that batch normalization is beneficial for backpropagation procedures.

4.11.1. Arc Tangent Function

When the upper and lower thresholds do not satisfy the conditions offered by tanh, sigmoid and softsign, another sigmoidal, the arc tangent function, "arctan", can be used, which offers half of π symmetrically to the origin as thresholds of $f(x)$ (Aghdam and Heravi, 2017). It can be used as a normalizer like other sigmoidals. The shape of it is shown in Figure 16 and given in Equation (24):

$$f(x) = tan^{-1}(x), \tag{24}$$

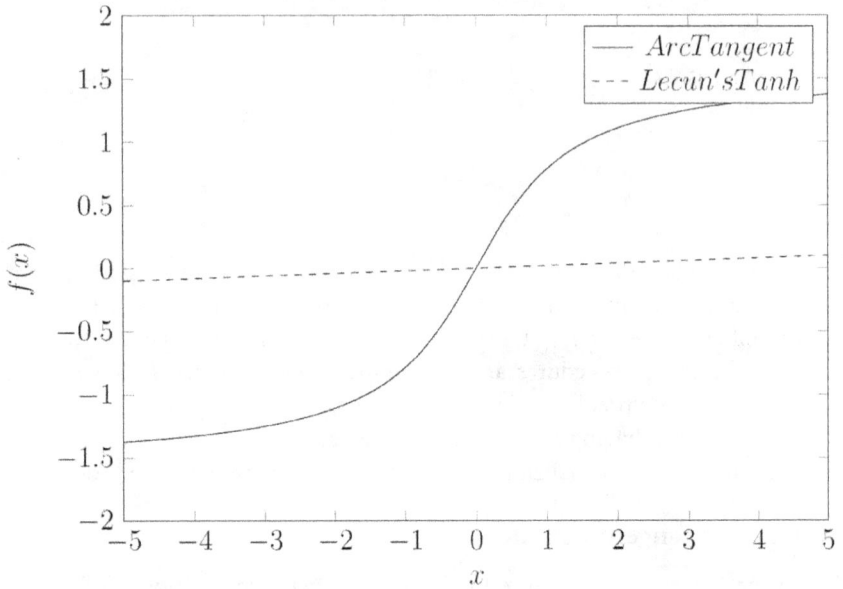

Figure 16: Arc Tangent Function.

In the range of $\left(-\dfrac{\pi}{2}, \dfrac{\pi}{2}\right)$, continuous in C^{∞}, monotonic for itself.

4.11.2. Lecun's Hyperbolic Tangent Function

An experimental activation function from Lecun can be used as a normalizer again with other sigmoidals. The shape of it is shown in Figure 16 and given in Equation (25):

$$f(x) = 1.7159 \tanh\left(\frac{2}{3}x\right),\tag{25}$$

In the range of $\left(-\dfrac{\pi}{2}, +\dfrac{\pi}{2}\right)$, continuous in C^{∞}, monotonic for itself.

4.12. Complementary log-log Function

Complementary log-log function which is the inverse of the cumulative density function (c.d.f.) of the extreme value (or log-Weibull) distribution and is used to compute hazard ratios is quite similar to the sigmoid family. It again provides the values between 0 and 1, but the results taken from an activation function as output can be interpreted as the hazard effect of reverse extreme value errors. Gomes and Ludermir (2008) point out that it outperforms logit and tanh activation functions according to mean squared errors of multi-layer perceptron networks. Cloglog, for short, is given in Figure 17 and is given in Equation (26).

$$f(x) = 1 - exp(-exp(x)),$$ (26)

in the range of $(0, +1)$, continuous in C^∞, monotonic for itself.

In general, the inverse cloglog version is commonly used as an activation function which solves classification problems.

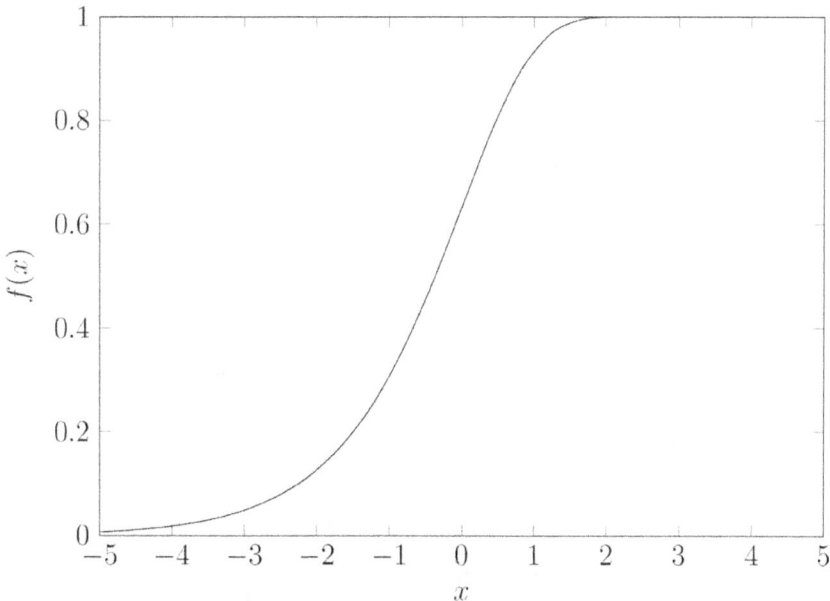

Figure 17: Complementary Log-Log Function.

4.13. Softplus Function

Since Glorot et al. (2011) emphasize the significance of softplus function and Zheng et al. (2015) prove the improvements in deep neural networks (deep learning) by using softpluses, it is used a new rectifier rather than using $x->max(0,x)$, that is ReLU, by smoothing its negative values. This smooth rectifier is given below in Equation (27) and its graph is presented in Figure 18:

$$f(x) = \ln(1 + e^x),\qquad\qquad(27)$$

in the range of $(0, +\infty)$, continuous in C^∞, monotonic for itself and its derivative.

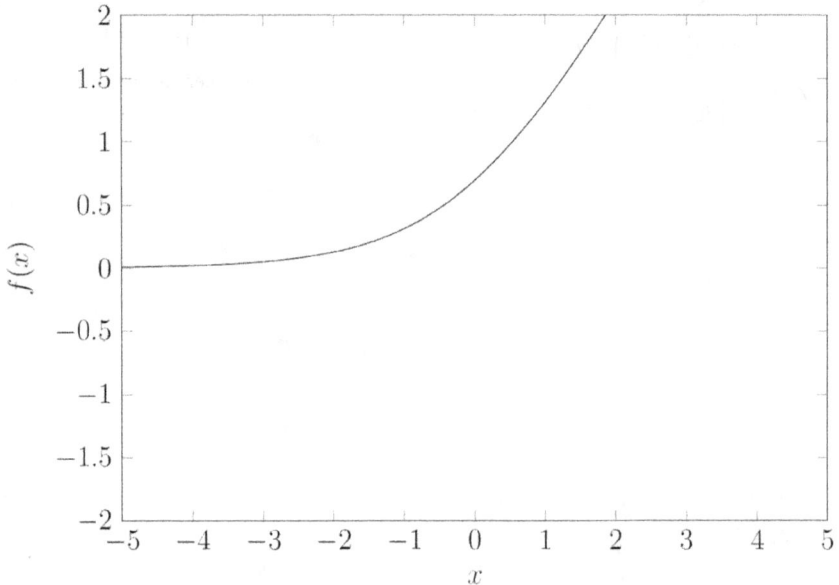

Figure 18: Softplus Function.

4.14. Bent Identity Function

After they were first defined in 1960s, Bent functions were published for the first time by Rothaus (1976). At the same time, Bent identifiers were being used for cryptography in USSR by V. A. Eliseev and O. P. Stepchenkov (Tokareva, 2015).

Bent functions can be classified in the Boolean function family (Çeşmelioğlu et al., 2016; Savický, 1994). Bent identity function is given below in Equation (28) and is visualized in Figure 19:

$$f(x) = \frac{\sqrt{x^2 + 1} - 1}{2} + x, \tag{28}$$

in the range of $(-\infty, +\infty)$, continuous in C^∞, monotonic for itself and its derivative.

Bent identity is an interesting function, it smoothly translates negative values of inputs to higher values and positive values of inputs to lower values. Nowadays, at the time of writing this study, it is used for solving Ethereum crypto-money problems if nonlinearity is needed to solve problems and cannot be solved by using other rectifiers.

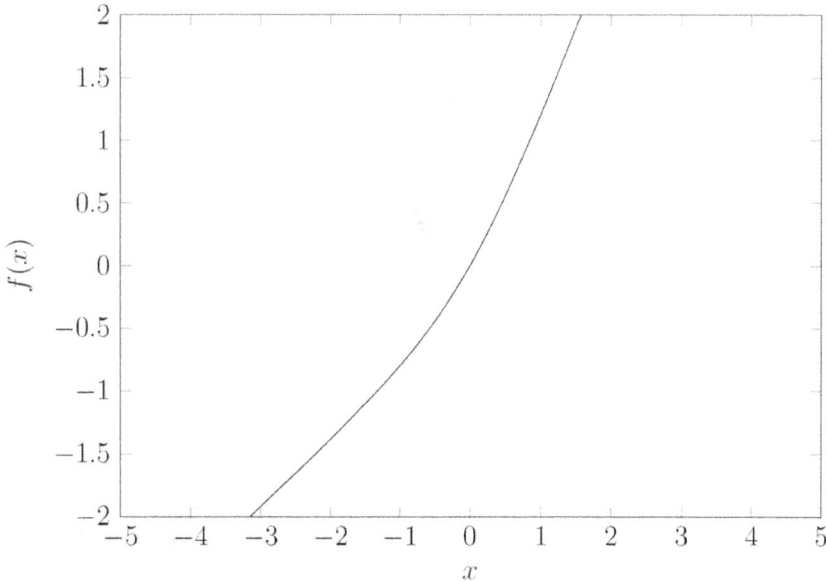

Figure 19: Bent Identity Function.

4.15. Soft Exponential Function

Soft exponential function as given in Equation (29) is recently proposed by Godfrey and Gashler (2015) who assert that the main property of SoftE is well-suited activation function for neural networks without empirical evidences. Soft exponential functions according to some α levels are drawn in Figure 20:

$$f(a,x) = \begin{cases} -\dfrac{\ln\left(1 - a(x + a)\right)}{a} & for\ a < 0 \\ x & for\ a = 0 \\ \dfrac{e^{ax} - 1}{a} + a & for\ a > 0 \end{cases} \qquad (29)$$

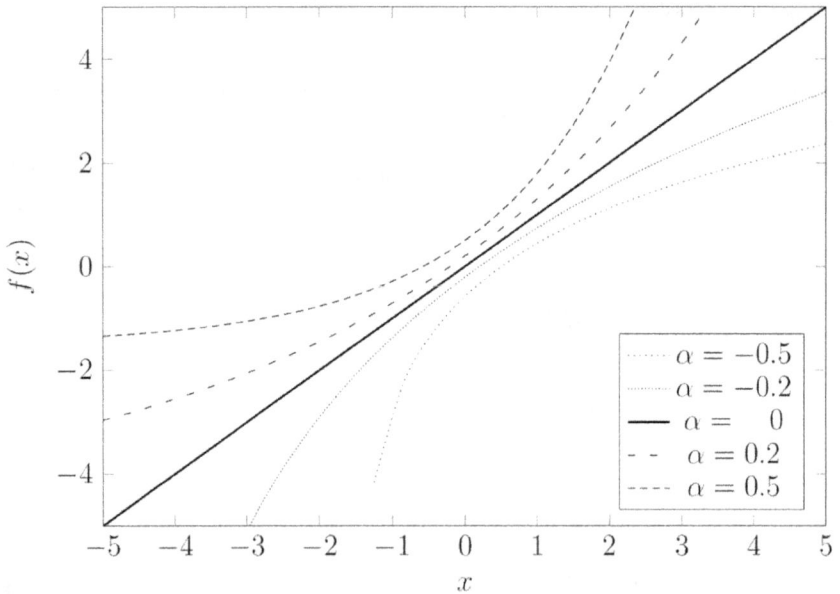

Figure 20: Soft Exponential Function.

in the range of $(-\infty, +\infty)$, continuous in C^∞, monotonic for itself and its derivative. Godfrey and Gashler (2015) present SoftE as a function that harmonizes the characteristics of logarithmic, linear and exponential functions since its multiplication can be done in logarithmic space and that is easily computable, differentiable and parameterized in the needs of neural network structure. Another strong property is that SoftE does not suffer from vanishing problem when training period is run.

5. Periodic Activation Functions

Activation functions covered so far are well suited for the neural networks that have a stable architecture that is well known and frequently used. These type of ANNs include a certain activation layer which is not varied by other neurons. Another architecture type of ANNs is named as adaptive artificial neural networks which consist of at least one additive neurons that affect an activation layer producing one or more outputs simultaneously. These classes of architectures need periodic activation functions where their derivatives result in oscillated, unstable outputs. This is only because adaptive neural networks contain both the weights of inputs and the learning rate of network (Widrow and Lehr, 1993). It may also change errors by adding recursively to the model.

5.1. Sinusoidals

Basic differentiation of a sinusoidal function compared to other types of functions comes from the existence of a repetition of oscillations in a certain period that can be modeled with a sine function and its derivatives. It does not matter whether these oscillations represent the same amplitude or not.

5.1.1. Sine Wave Function

Sine wave[9] is a geometric form of wave oscillating periodically with a certain amplitude looking like S-shaped wave around the x-axis in the range of -1 and +1 (Parascandolo et al., 2016). It is called and abbreviated as "sine" given in Equation (30) which results in its first derivative as cosine, but first derivative of cosine is minus sine.

$$f(x,t) = A\sin(kx \pm \omega t + \varphi) + D \tag{30}$$

in the range of $(-\infty, +\infty)$, continuous in C^∞, t is the oscillation period, (-) sign is for right moves, (+) sign is for left moves.

The main and most common use of sine wave functions as illustrated in Figure 21 is signal processing. Since sine or cosine are non-quasiconvex

---

9   The first song of Mogwai's "Rock Action" album that is revolutionary one in post-rock genre.

functions and therefore non-monotonic, it is useful if an increase or decrease in the correlation or dependency between activation and input oscillate the activation function in the learning process. This may easily cause the learning process to derail if the number of epochs increases or learning rate is higher than its optimal level; in addition to this, this can cause a decrease in the speed rate of learning, gradient vanishing or converging local minima -rather than global – problem as Lapedes and Farber (1987) stated. In recent studies, it is shown that well-known periodic activation functions such as sine may easily suffer from high ram swelling issue, however, sine performs better than other monotonic functions in Recurrent Neural Networks (RNN) while predicting in short term (Sopena and Alquezar, 1994; Alquezar and Sanfeliu, 1994). Another property of sine and cosine is the averages of them are zero, which is useful property especially for statisticians.

5.1.2. Cardinal Sine Function (Sinc)

Cardinal sine function, abbrevated as sinc, is the candidate for the weirdest activation function among both monotonic and periodic ones. Şelariu et al. (2014) state the concept of cardinality as "a number equal to a number of finite aggregate" and classify it in centric functions because it is centered on the x-axis while waving around it, like a Mexican hat. In general, as given in Equation (31), it is normalized by using π as Equation (32), both of which them are shown in Figure 22.

$$f(x) = \frac{\sin(x)}{x} \tag{31}$$

and normalized version of Equation (31),

$$f(x) = \frac{\sin(\pi^* x)}{\pi^* x} \tag{32}$$

in the range of $(-\infty, +\infty)$, continuous in C^∞.

The most interesting point for the normalization of the *sinc* activation function is that it can result in a Fourier transform of a rectangular function that may be used for getting 1 for determined interval of x values like rectifier units.

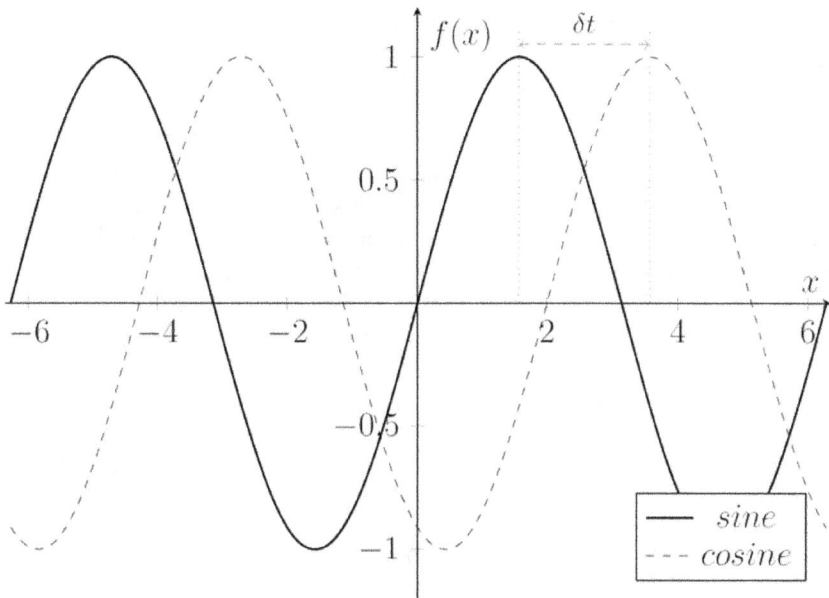

Figure 21: Sine and Cosine Wave Functions.

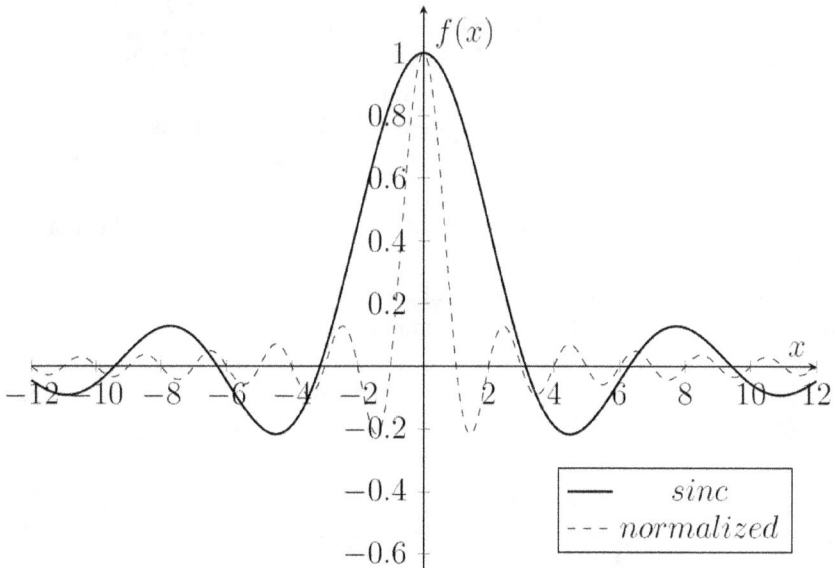

Figure 22: Sinc Function.

5.1.3. Fourier Transform (FT, DFT/FFT)

There are three concepts that must be explained respectively. The first one is fourier analysis, that area is a way to represent functions as an exact or approximated version by means of summing certain trigonometric functions. Secondly, the Fourier series in the Fourier analysis are generally related to making a representation by simply adding sine waves to each other. Fourier transformation, however, by using fourier series, converts a series to an orthogonal one into new space with its linear form (Bracewell and Bracewell, 1986). This transformation also helps state differential equations as linearized. A continuous fourier transformation is described in Equation (33) and (34) and is visualized in Figure 23.

$$\hat{f}(k) = \int_{-\infty}^{\infty} f(x) e^{-2\pi i x k} dx, \tag{33}$$

$$f(k) = \int_{-\infty}^{\infty} \hat{f}(k) e^{2\pi i k x} dk, \tag{34}$$

in the range of $(-\infty, +\infty)$.

Fourier transformation has some properties which are linearity, frequency and time shifting, time scaling, which are useful when they are processed as activation functions. Additionally, the discrete version (DFT) which is not described and shown can be used for image processing, sound waves if the data are finite or the processing of them are relatively trivial because it helps to transform it easily by metamorphosing into a fast fourier transform (FFT) (Wickerhauser, 1994).

Meaning of fourier transform literally is the sum over all time of the inputs of $f(x)$ multiplied by an exponential and the result is the Fourier coefficients $\hat{f}(k)$, when these coefficients are multiplied by a sinusoid of appropriate frequency k, they yield the constituent sinusoidal component of the original inputs.

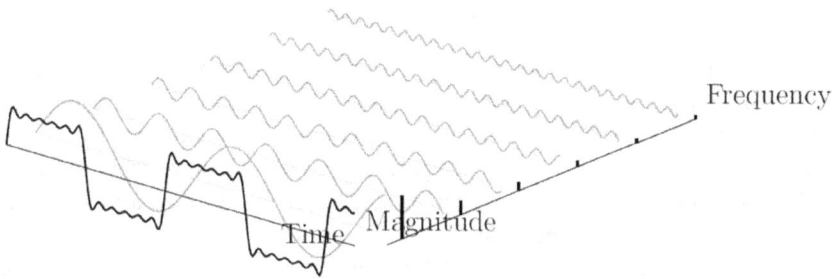

Figure 23: Fourier Transform for Continuous Series.

For the relationship between Equation (33) and (34) should be denoted as $f(x) \rightarrow \hat{f}(k)$ meaning "Fourier transform" of these two equations mapped, and $\hat{f}(k) \rightarrow f(x)$, meaning "Inverse fourier transform" of the equations mapped. This mapping procedure is abbreviated with capital letters such as FT and IFT.

5.1.4. Discrete-time Dimensional Fourier Transform (DTFT, Shannon-Nyquist)

Discrete-time dimensional Fourier transform (DTFT) having time domain where y-axis contains ampitudes of inputs (signals) is another version of Fourier Transform. There are two types of DTFT, which are periodic and aperiodic according to Memon et al. (2014), but in general, aperiodic ones are relatively hard to model and converted into a serie. DTFT is given in Equation (35) and

$$X_{2\pi}(\omega) = \sum_{n=-\infty}^{\infty} x[n]e^{-i\omega n} \tag{35}$$

$$X_{1/T}(f) = X_{2\pi}(2\pi fT)\underline{\underline{\text{def}}} \sum_{n=-\infty}^{\infty} \underbrace{T.x(nT)}_{x[n]}e^{-i2\pi fTn}$$

$$\overset{Poisson\,f.}{=.} \sum_{n=-\infty}^{\infty} X(f-k/T), \tag{36}$$

in the range of $(-\infty, +\infty)$.

5.1.5. Short-Time Fourier Transform (Gabor, STFT)

Short-time or short-term fourier transform (STFT), in general, divides whole time into small parts which are equal, afterwards, for each part, fourier transforms are calculated in order to determine sinusoidal frequencies and are respectively bound to see all local changes of each part. The concept of short-term indicates each part or phases that were separated into, so it can be called a mix of time dimensional and fourier transform.

Kalayci and Ozdamar (1995) regard it as a very beginning version of wavelets for an activation function and assert that it has some drawbacks such as Heisenberg's Uncertainty Principle stated as inequality: $\text{time*bandwidth} = \Delta f \Delta t \geq \dfrac{1}{4\pi}$. It indicates neither times nor bandwith may be jointly small. In general, STFT is used to analyse audio signals especially those which have harmonic patterns (Wu and Liu, 2008). It also fails when the data are nonstationary which causes arising asymmetries between phases or parts.

STFT has two sub-categories according to its time dimension, it can be discrete or continuous. Discrete version of STFT is given below in Equation (37), where the data must be split into frames:

$$STFT\left\{x[n]\right\}(m,\omega) = \sum_{n=-\infty}^{\infty} x[n]w[n-m]e^{-jwn} \tag{37}$$

In continuous case, a window function recalled Gaussian window that centered around zero and expected value of that is zero for multiplying it for short-terms is needed again with an R parameter. R parameter is a weight for time axis as given in Equation (38).

$$\sum_{n=-\infty}^{\infty} x(n)w(n-mR)e^{-jwn} \tag{38}$$

The dependency between discrete and continuous STFT is given below:

$$DSTFT_{\omega}\left(x \cdot SHIFT_{mR}(w)\right), \tag{39}$$

in the range of $(-\infty, +\infty)$.

In Equation (39), shift in time axis is the R parameter that can be determined exogenously or endogenously.

5.1.6. Wavelet Transform

Wavelet is seen in the compression of images with JPEG2000 format, thanks to decreasing the lengths of inputs of images to half of it, so size of images could be decreased. Basically, when Fourier says frequencies of inputs, wavelet indicates both frequencies and time, in econometric terms, wavelet is actually a panel data version of the fourier series. Differences between all fourier variants and wavelets are shown in Figure 24.

According to Hippenstiel (2001), wavelets have one dimension for time and second dimension for transform. If both inputs and outputs are continuous, then it is called continuous wavelet transform (CWT), the most common transform of wavelets (Ombao et al., 2016) used as an activation function (Pati and Krishnaprasad, 1993). If the outputs are discrete while inputs are continuous, it is called discrete wavelet transform (DWT). The functions for them are given below respectively:

$$X_w(a,b) = \frac{1}{\sqrt{|a|}} \int_{-\infty}^{\infty} x(t) \psi^* \left(\frac{t-b}{a} \right) dt \tag{40}$$

in the range of $(-\infty, +\infty)$.

The wavelet coefficients for continuous series of transformation are computed as following:

$$C_{jk} = \left[W_\psi f \right] \left(2^{-j}, k2^{-j} \right), \tag{41}$$

in the range of $(-\infty, +\infty)$.

Furthermore, DWT is as given below:

$$\psi(x) = \sum_{k=-\infty}^{\infty} (-1)^k a_{N-1-k} \psi(2x - k), \tag{42}$$

in the range of $(-\infty, +\infty)$.

In addition to these, there are two other types of wavelet transforms (Hippenstiel, 2001) which are not described in detail: If inputs are given with time detail with discrete form, it is called discrete-time wavelet transform (DTWT) but outputs of transform are also discrete. Besides, if outputs transformed are continuous, then it is called discrete-time continuous wavelet transform (DTCWT). Mostly, continuous cases are harder to implement than discrete ones, the problems faced may be turned into discrete case and then be solved. It is less costly in the implementation period for both processing and computing, which are the most common problems in machine learning if they are used as an activation function that is vulnerable and sensible to the vanishing gradient problem that is not desired.

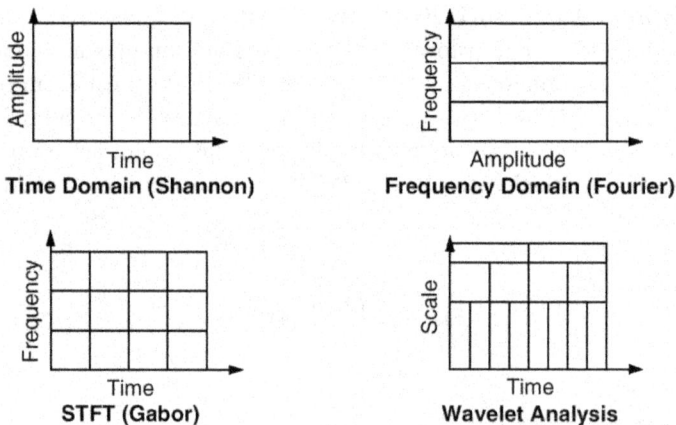

Figure 24: Fourier and its variants by domain.

5.2. Non-sinusoidals

As it can be understood from the title, in this section, certain activation functions which do not include any sinusoidal function are covered. The functions classified into Non-sinusoidal that can be periodic or not, it is not a must but these functions can be expanded into periodic.

5.2.1. Gaussian (Normal) Distribution Function

The nature of Gaussian which might be the oldest function in the history of statistics (Stigler, 1986) comes from its nonperiodicity, since it produces a reference distribution at once. It does not need to explain its reputation in statistics.

But it is required that periodicity, a property that can be applied to most non-monotonic functions, can be introduced by adding a periodic pattern to most functions. The nature provides that the properties of everything living or existing in it are normally distributed if these properties are eventually found when time dimension increases. Thank to Laplace, the central limit theorem proves that assertion when the number of observations (n) increases (Bishop, 2006). So, if a distribution which has less observations about the information tends to turn into Gaussian distribution if n increases. Lin et al. (2004) firstly use Gaussian as a regularization parameter of estimation. The first of periodic Gaussian is shown in Figure 25 and are given in Equations(43 and 44):

$$f\left(x|\mu,\sigma^2\right)=\frac{1}{\sqrt{2\pi\sigma^2}}e^{-\frac{(x-\mu)^2}{2\sigma^2}},\tag{43}$$

in the range of $\left(-\infty,+\infty\right)$ and continuous in C^∞.

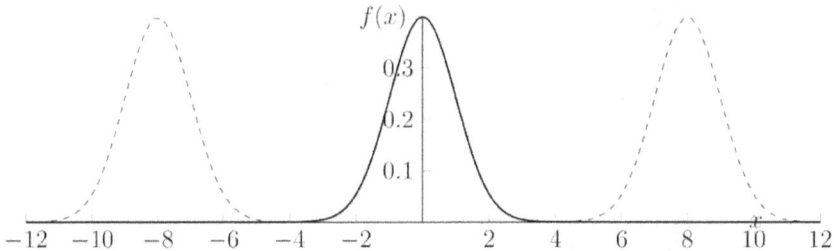

Figure 25: Periodic Gaussian Function.

The periodicity of Gaussian or normal distribution function can be given by implementing the following equation in the constrained space:

$$p(x)=f\left(\left(\left(x+\frac{N}{2}\right)mod\ N\right)-\frac{N}{2}\right)\tag{44}$$

On the other hand, by definition, wrapped normal distribution is another kind of normal distribution which wraps around a unit circle and is widely used in directional statistics, another periodic version of normal distribution function is called Von Mises distribution function that approximates a wrapped one.

5.2.2. Square Wave Function

Square wave can be thought of as an expanded version of the Heaviside step function by implementing its property of periodicity. It is also an application of binary transmission of a wave from a base level to the desired level. Effects of electric guitars, such as overdrive and distortion effects, are examples of a square wave function. Orthogonality of waves are required according to Thompson et al. (2001) for square waves. The function that produces square waves is given below in Equations (45) and (46) representing the resemblance to the Heaviside unit function if it is denoted as $u(t)$ and is illustrated in Figure 26.

$$x(t) = sgn\big(\sin[t]\big), v(t) = sgn\big(\cos[t]\big) \tag{45}$$

in the range of $(-\infty, +\infty)$ and continuous in C^∞.

$$x(t) = \sum\nolimits_{n=-\infty}^{\infty} \Pi(t - nT) = \sum\nolimits_{n=-\infty}^{\infty} \left(u\left[t - nT + \frac{1}{2}\right] - \left[t - nT - \frac{1}{2}\right] \right) \tag{46}$$

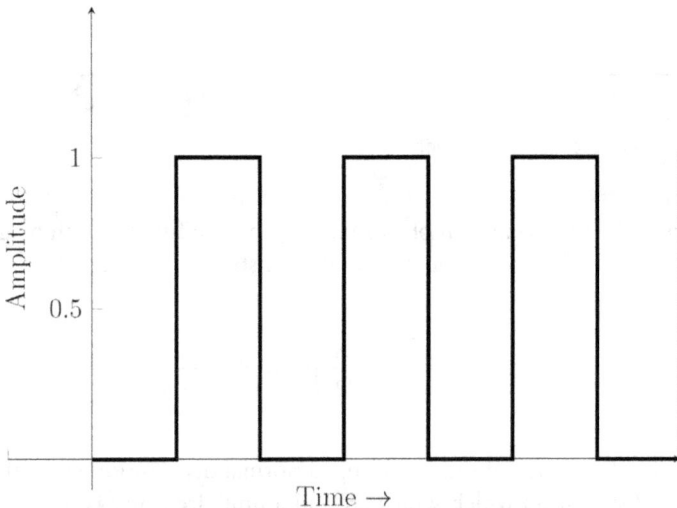

Figure 26: Square Wave Function.

5.2.3. Triangle Wave Function

The shape of this non-sinusoidal function gives it its name as it draws small triangles around the x-axis (Tansel et al., 1991). As Foresee and Hagan (1997) used a triangle wave function to estimate Gauss-Newton approximation to Bayesian regularization (GNBR) algorithm for a three real world problem with application of a univariate regression, time series estimation and an artificial chaotic series model regardless of its contents. They generate the lowest costs by using triangle waves for artifical neural networks including a regularization that improves the results leading to significant decrease in errors. The functional form of the triangle wave function is stated in Equation (47) and is represented in Figure 27.

$$x(t) = \frac{2}{a}\left(t - a\left[\frac{t}{a} + \frac{1}{2}\right]\right)(-1)^{\left[\frac{t}{a} + \frac{1}{2}\right]} \tag{47}$$

in the range of $(-\infty, +\infty)$ and continuous in C^{∞}, where a is the amplitude of triangle wave.

The integral of a square wave is equivalent to a triangle wave as follows:

$$\int sgn(\sin(x))\, dx \tag{48}$$

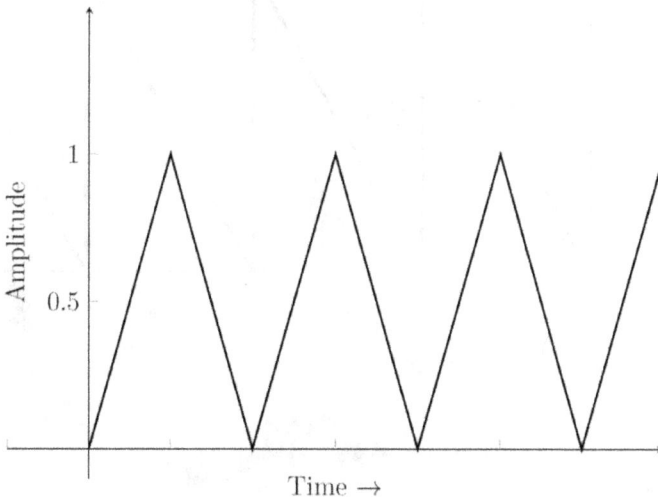

Figure 27: Triangle Wave Function.

5.2.4. Sawtooth Wave Function

Sawtooth wave is quite similar to a triangle wave function as it can be seen in Figure 28. Bose (2007) generates a 1-5-3 neural network model that converges and fits the data properly, that is, it minimizes the errors by using sawtooth waves for power electronics and motor drives. Wang et al. (2017) use it in the cellular neural network to identify the patterns of HIV carriers by elaborating the data from images of liver tissues. Sawtooth waves can be drawn as Figure 28. In the Equation (49), the absolute value of a sawtooth wave function may result in a triangle wave as follows:

$$x(t) = 2\left(\frac{t}{a} - \left[\frac{t}{a} + \right]\right) \tag{49}$$

in the range of $(-\infty, +\infty)$ and continuous in C^∞.

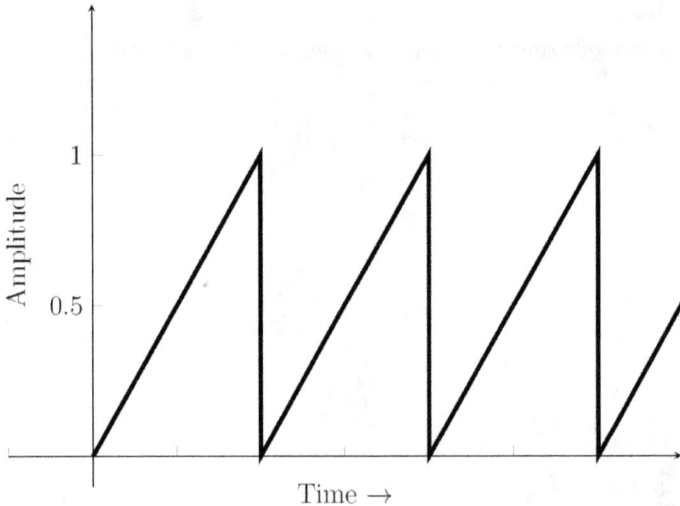

Figure 28: Sawtooth Wave Function.

5.2.5. S-shaped Rectifed Linear Unit (SReLU)

As introduced first by Jin et al. (2016a), it is designed to learn both convex and non-convex functions by imitating some functions in physics. Jin et al. (2016a) state that it contains three linear functions which are reshaped with four parameters leading to two knuckles (t_l and t_r for Equation (51)). Since the initial parameters are unknown at the first stage, Jin et al. (2016b) and Jin et al. (2016a) propose it can be manually and exogeneously determined. But if it is set to too large or too small, SReLu may not work well. To deal with this problem, they propose:

- Initialize each o_i to be $(t_i, 0, a_i)$ in all layers where t_i is any positive real number and $a_i \in (0,1)$, and it is freezed the update of the parameters of SReLU during the initial several training epochs.
- SReLU is degenerated into a conventional LReLU at the beginning of the training. Then upon the end of the freezing phase, it is set t_i^r to be the largest k_{th} value of each SReLU's input from all training data, i.e

$$t_i^r = supp(X_i, k);$$
(50)

where $supp(X_i, k)$ calculates the k th largest value from the set X, and X_i represents all the input values of an individual SReLU.

SReLu is explained with Equation (51) given below:

$$f_{t_l, a_l, t_r, a_r}(x) = \begin{cases} t_l + a_l(x - t_l) & for\, x \leq t_l \\ x & for\, t_l < x < t_r \\ t_r + a_r(x - t_r) & for\, x \geq t_r \end{cases}$$
(51)

in the range of $(-\infty, +\infty)$ and continuous in C^0. The parameters t_l, a_l, t_r, a_r which are learnable parameters are predetermined or endogeneously determined, and can be defined in real numbers.

The first drawback, as stated previously, is that it is too hard to determine its initial parameters. Second drawback is its differentiability except its two points serving three piecewise linear functions to merge, so at these points, it is impossible to differentiate SReLu. Besides, Jin et al. (2016a) found that SReLu performs so well than other rectifiers when solving deep neural networks.

5.2.6. Adaptive Piecewise Linear Unit (APLU)

While the term adaptive mainly describes the non-monotonicity of this activation function, APLU has two terms, the first one includes a ReLU, but the second term plays a role as a regularization parameter in gradient descent. Agostinelli et al. (2014) propose a new kind of activation unit that learns from neurons independently by using gradient descent, which is the most distinctive property rather than other rectifier units. Since, practically, any piecewise-linear function that must be continuous can be denoted as APLU by accepting the learning parameter a_i in Equation (52) as zero; Agostinelli et al. (2014) assert ApLU is a generalization of piecewise-linear function. Due to the regularization parameter, APLU can have more than one hinge, this property may turn it into a periodic function which cannot satisfy monotonicity conditions. APLU is given in Equation (52):

$$f(x) = \max(0,x) + \sum_{s=1}^{s} a_i^s \max(0, -x + b_i^s) \qquad (52)$$

in the range of $(-\infty, +\infty)$.

  APLU can learn from each neuron as their weights, so each iteration in a machine learning or deep learning process (the number of epochs) gradually shapes and provides a neuron-specific adaptive piecewise activation function. Agostinelli et al. (2014) assert that APLU significantly increases success of prediction in CIFAR-10 and CIFAR-100 experiments. But, since there is no consistency for an activation function for each neuron, it may lead to increased fragility of the complete system if there exists at least one failure in a neuron. In the next step, it may trigger to produce bias for other neurons.

6. Bias Unit

A bias unit is an "extra" neuron added to each pre-output layer with a value of 1. Bias units are not connected to any previous layers and in this sense do not represent a true "activity".

The bias units are characterized by the text "+1". As it can be seen, a bias unit is just appended to the start/end of the input and each hidden layer, and is not inflated by the values in the previous layer. In other words, these neurons do not have any incoming connections.

References

Aghdam, H. H. and Heravi, E. J. (2017). Guide to convolutional neural networks. New York, NY: Springer. doi, 10:978–3.

Agostinelli, F., Hoffman, M., Sadowski, P., and Baldi, P. (2014). Learning activation functions to improve deep neural networks. arXiv preprint arXiv:1412.6830.

Alquezar, R. and Sanfeliu, A. (1994). A hybrid connectionist-symbolic approach to regular grammatical inference based on neural learning and hierarchical clustering. In Grammatical Inference and Applications, pages 203–211. Springer Berlin Heidelberg.

Batres-Estrada, B. (2015). Deep learning for multivariate financial time series.

Bishop, C. M. (2006). Pattern recognition and machine learning, Springer.

Bose, B. K. (2007). Neural network applications in power electronics and motor drives—an introduction and perspective. IEEE Transactions on Industrial Electronics, 54(1):14–33.

Bracewell, R. N. and Bracewell, R. N. (1986). The Fourier transform and its applications, volume 31999. McGraw-Hill New York.

Çeşmelioğlu, A., Meidl, W., and Pott, A. (2016). There are infinitely many bent functions for which the dual is not bent. IEEE Transactions on Information Theory, 62(9):5204–5208.

Clevert, D.-A., Unterthiner, T., and Hochreiter, S. (2015). Fast and accurate deep net- work learning by exponential linear units (elus). arXiv preprint arXiv:1511.07289.

Cox, D. R. (1992). Regression models and life-tables. In Breakthroughs in statistics, pages 527–541. Springer.

Dai, W., Yang, Q., Xue, G.-R., and Yu, Y. (2007). Boosting for transfer learning. In Proceedings of the 24th international conference on Machine learning, pages 193–200. ACM.

Dietterich, T. G. (2000). An experimental comparison of three methods for constructing ensembles of decision trees: Bagging, boosting, and randomization. Machine learning, 40(2):139–157.

Elliott, D. L. (1993). A better activation function for artificial neural networks. Cite- seerX.

Foresee, F. D. and Hagan, M. T. (1997). Gauss-newton approximation to bayesian learning. In Neural networks, 1997., international conference on, volume 3, pages 1930–1935. IEEE.

Friedman, J., Hastie, T., and Tibshirani, R. (2001). The elements of statistical learning, volume 1. Springer series in statistics New York.

Glorot, X., Bordes, A., and Bengio, Y. (2011). Deep sparse rectifier neural networks. In Proceedings of the Fourteenth International Conference on Artificial Intelligence and Statistics, pages 315–323.

Godfrey, L. B. and Gashler, M. S. (2015). A continuum among logarithmic, linear, and exponential functions, and its potential to improve generalization in neural networks. In Knowledge Discovery, Knowledge Engineering and Knowledge Management (IC3K), 2015 7th International Joint Conference on, volume 1, pages 481–486. IEEE.

Gomes, G. S. d. S. and Ludermir, T. B. (2008). Complementary log-log and probit: activation functions implemented in artificial neural networks. In Hybrid Intelli- gent Systems, 2008. HIS'08. Eighth International Conference on, pages 939–942. IEEE.

Goodfellow, I., Bengio, Y., and Courville, A. (2016). Deep learning. MIT press. Goodfellow, I. J., Warde-Farley, D., Mirza, M., Courville, A., and Bengio, Y. (2013). Maxout networks. arXiv preprint arXiv:1302.4389.

Greene, W. H. (2003). Econometric analysis. Pearson Education India.

Haykin, S. S. (2001). Neural networks: a comprehensive foundation. Tsinghua University Press.

Haykin, S. S. (2009). Neural networks and learning machines, volume 3. Pearson Upper Saddle River, NJ, USA:.

Hegel, G. W. F. (1861). Lectures on the Philosophy of History. Henry G. Bonn. Hinton, G. E., Srivastava, N., Krizhevsky, A., Sutskever, I., and Salakhutdinov, R. R. (2012). Improving neural networks by preventing co-adaptation of feature detectors. arXiv preprint arXiv:1207.0580.

Hippenstiel, R. D. (2001). Detection theory: applications and digital signal processing. CRC Press.

Jain, A. K., Mao, J., and Mohiuddin, K. M. (1996). Artificial neural networks: A tutorial. Computer, 29(3):31–44.

James, G., Witten, D., Hastie, T., and Tibshirani, R. (2013). An introduction to statistical learning, volume 112. Springer.

Jin, X., Xu, C., Feng, J., Wei, Y., Xiong, J., and Yan, S. (2016a). Deep learning with s-shaped rectified linear activation units. In AAAI, pages 1737–1743.

Jin, X., Yuan, X., Feng, J., and Yan, S. (2016b). Training skinny deep neural networks with iterative hard thresholding methods. arXiv preprint arXiv:1607.05423.

Kalayci, T. and Ozdamar, O. (1995). Wavelet preprocessing for automated neural net- work detection of eeg spikes. IEEE engineering in medicine and biology magazine, 14(2):160–166.

Klambauer, G., Unterthiner, T., Mayr, A., and Hochreiter, S. (2017). Self-normalizing neural networks. arXiv preprint arXiv:1706.02515.

Lapedes, A. and Farber, R. (1987). Nonlinear signal processing using neural networks: Prediction and system modelling.

Lecun, Y., Bottou, L., Bengio, Y., and Haffner, P. (1998). Gradient-based learning applied to document recognition. Proceedings of the IEEE, 86(11):2278–2324.

LeCun, Y. A., Bottou, L., Orr, G. B., and Mu¨ller, K.-R. (2012). Efficient backprop. In Neural networks: Tricks of the trade, pages 9–48. Springer.

Lin, Y., Brown, L. D., et al. (2004). Statistical properties of the method of regularization with periodic gaussian reproducing kernel. The Annals of Statistics, 32(4):1723–1743.

Maas, A. L., Hannun, A. Y., and Ng, A. Y. (2013). Rectifier nonlinearities improve neural network acoustic models. In Proc. ICML, volume 30.

McCorduck, P. (1979). Machines who think: A personal inquiry into the history and prospects of artificial intelligence. Wh freeman San Francisco.

McMahan, H. B., Moore, E., Ramage, D., Hampson, S., et al. (2016). Communication- efficient learning of deep networks from decentralized data. arXiv preprint arXiv:1602.05629.

Memon, A. P., Uqaili, M. A., Memon, Z. A., and Tanwani, N. K. (2014). Time-frequency and artificial neural network applications and analysis for electrical system power quality disturbances in matlab. International Journal of Innovative Technology and Exploring Engineering (IJITEE), 3.

Mitchell, T. M. (1997). Machine learning. 1997. Burr Ridge, IL: McGraw Hill, 45(37):870–877.

Nair, V. and Hinton, G. E. (2010). Rectified linear units improve restricted boltzmann machines. In Proceedings of the 27th international conference on machine learning (ICML-10), pages 807–814.

Ng, A. Y., Coates, A., Diel, M., Ganapathi, V., Schulte, J., Tse, B., Berger, E., and Liang, E. (2006). Autonomous inverted helicopter fl t via reinforcement learning. In Experimental Robotics IX, pages 363–372. Springer.

Ombao, H., Lindquist, M., Thompson, W., and Aston, J. (2016). Handbook of Neuroimaging Data Analysis. CRC Press.

Osher, S. and Fedkiw, R. (2003). Implicit functions. In Level Set Methods and Dynamic Implicit Surfaces, pages 3–16. Springer.

Panicker, M. and Babu, C. (2012). Efficient fpga implementation of sigmoid and bipolar sigmoid activation functions for multilayer perceptrons. IOSR Journal of Engineering, pages 1352–1356.

Parascandolo, G., Huttunen, H., and Virtanen, T. (2016). Taming the waves: sine as activation function in deep neural networks.

Pati, Y. C. and Krishnaprasad, P. S. (1993). Analysis and synthesis of feedforward neural networks using discrete affine wavelet transformations. IEEE Transactions on Neural Networks, 4(1):73–85.

Ploskas, N. and Samaras, N. (2016). GPU Programming in Matlab. Morgan Kaufmann. Rice, H. G. (1953). Classes of recursively enumerable sets and their decision problems. Transactions of the American Mathematical Society, 74(2):358–366.

Rothaus, O. S. (1976). On "bent" functions. Journal of Combinatorial Theory, Series A, 20(3):300–305.

Royden, H. L. and Fitzpatrick, P. (1988). Real analysis. Number 8. Macmillan New York.

Russell, S., Norvig, P., and Intelligence, A. (1995). A modern approach. Artificial Intelligence. Prentice-Hall, Egnlewood Cliffs, 25:27.

Sauri, J. (1774). Cours complet de math´ematiques, volume 1. Ruault.

Savicky, P. (1994). On the bent boolean functions that are symmetric. European Journal of Combinatorics, 15(4):407–410.

Şelariu, M. E., Smarandache, F., and Ni¸tu, M. (2014). Cardinal functions and integral functions. Collected Papers, V, page 45.

Sopena, J. and Alquezar, R. (1994). Improvement of learning in recurrent networks by substituting the sigmoid activation function. In ICANN, volume 94, pages 417–420.

Standage, T. (2002). The Turk: The life and times of the famous eighteenth-century chess-playing machine. Walker & Company.

Stigler, S. M. (1986). The history of statistics: The measurement of uncertainty before 1900. Harvard University Press.

Tansel, I., Wagiman, A., and Tziranis, A. (1991). Recognition of chatter with neural networks. International journal of machine tools and manufacture, 31(4):539–552.

Thompson, A. R., Moran, J. M., and Swenson, G. W. (2001). Interferometry and synthesis in radio astronomy.

Tokareva, N. (2015). Bent functions: results and applications to cryptography. Academic Press.

Trottier, L., Gigu`ere, P., and Chaib-draa, B. (2016). Parametric exponential linear unit for deep convolutional neural networks. arXiv preprint arXiv:1605.09332.

Verhulst, P.-F. (1838). Notice sur la loi que la population suit dans son accroissement. correspondance math´ematique et physique publiee par a. Quetelet, 10:113–121.

Verhulst, P.-F. (1977). A note on the law of population growth. In Mathematical Demography, pages 333–339. Springer.

Wang, M., Min, L., Litscher, G., and Li, M. (2017). Dynamic analysis of coupled sawtooth and rectangle cellular neural networks with application in hbv patients' b-scan images. Integrative Medicine International, 4(1–2):19–30.

Wickerhauser, M. V. (1994). Adapted wavelet analysis from theory to software. IEEE press.

Widrow, B. and Lehr, M. A. (1993). Adaptive neural networks and their applications. International Journal of Intelligent Systems, 8(4):453–507.

Wu, J.-D. and Liu, C.-H. (2008). Investigation of engine fault diagnosis using dis- crete wavelet transform and neural network. Expert Systems with Applications, 35(3):1200–1213.

Xu, B., Wang, N., Chen, T., and Li, M. (2015). Empirical evaluation of rectified activations in convolutional network. arXiv preprint arXiv:1505.00853.

Zalta, E. N. et al. (2003). Stanford encyclopedia of philosophy.

Zeng, Z., Huang, T., and Zheng, W. X. (2010). Multistability of recurrent neural networks with time-varying delays and the piecewise linear activation function. IEEE Transactions on Neural Networks, 21(8):1371–1377.

Zheng, H., Yang, Z., Liu, W., Liang, J., and Li, Y. (2015). Improving deep neural networks using softplus units. In Neural Networks (IJCNN), 2015 International Joint Conference on, pages 1–4. IEEE.

LaTeX Applications

4. Monotonic Activation Functions

```
\begin{figure}[!htb]
    \centering
    \caption{Monotonic Functions.}
    \begin{tikzpicture}
    \label{fig.monotone}
    \begin{axis}[width=12cm,height=9cm,ylabel=$f(x)$,xlabel=$x$,ymin=-1,ymax=1,xmin=-
    5,xmax=5,
    legend pos= north west]
    \addplot[thick,smooth] {(1/(2+exp(-(0.9071982*x))))};
    \addlegendentry{$Nondecreasing, monotone function$}
    \addplot[dotted,thick,dash pattern={on 7pt off 2pt on 1pt off 3pt}] {(1/(0.2+exp((0.73*x))))};
    \addlegendentry{$Nonincreasing, monotone function$}
    \addplot[domain=-5:2*pi,samples=200,densely dashed,smooth]{sin(deg(x))}node[right,pos=
    0.9]{$f(x)=\sin x$};
    \addlegendentry{$Non-monotone (periodic) function$}
    \end{axis}
    \end{tikzpicture}
\end{figure}
```

4.1. Linear Function & Identity Function

```
\begin{figure}[!htb]
    \centering
    \caption{Linear Function.}
    \begin{tikzpicture}
    \label{fig.linear}
    \begin{axis}[width=12cm,height=9cm,ylabel=$f(x)$,xlabel=$x$,ymin=0,ymax=5,xmin=-
    5,xmax=5,
    legend pos= north west]
    \addplot[black,smooth] {0.75*x};
    \addlegendentry{$\alpha =0.75 (Linear)$}
    \addplot[dashed,smooth] {x};
    \addlegendentry{$\alpha =1.0 (Identity)$}
    \end{axis}
    \end{tikzpicture}
\end{figure}
```

4.1.2 Piecewise Linear Function

\begin{figure}[!htb]
    \centering
    \caption{Piecewise Linear Function.}
    \begin{tikzpicture}
    \label{fig.piecelinear}
    \begin{axis}[width=12cm,height=9cm,ylabel=$f(x)$,xlabel=x,ymin=0,ymax=1,xmin=-5,xmax=5,
    legend pos= north west]
    \addplot[thick, line width=1pt] coordinates
    {(-5,0) (-4,0) (3,1) (5,1)};
    \addlegendentry{$Piecewise Linear$}
    \addplot[] coordinates
    {(-5,0) (-4,0) (3,1) (5,1)};
    \addlegendentry{$\alpha_{min}=-4$}
    \addplot[] coordinates
    {(-5,0) (-4,0) (3,1) (5,1)};
    \addlegendentry{$\alpha_{max}=+3$}
    \end{axis}
    \end{tikzpicture}
\end{figure}

4.2 Threshold Function

\begin{figure}[!htb]
    \centering
    \caption{Threshold (Unit Heaviside / Binary / Step) Function.}
    \begin{tikzpicture}
    \label{fig.threshold}
    \begin{axis}[width=12cm,height=9cm,ylabel=$f(x)$,xlabel=x,ymin=0,ymax=2,xmin=-5,xmax=5]
    \addplot[thick, mark=x,line width=2pt] coordinates
    {(-5,0) (0,0) (0,1) (5,1)};
    \end{axis}
    \end{tikzpicture}
\end{figure}

4.2 Threshold Function

\begin{figure}[!htb]
    \centering
    \caption{Threshold (Unit Heaviside / Binary / Step) Function.}
    \begin{tikzpicture}
    \label{fig.threshold}

```
\begin{axis}[width=12cm,height=9cm,ylabel=$f(x)$,xlabel=$x$,ymin=0,ymax=2,xmin=-
5,xmax=5]
\addplot[thick, mark=x,line width=2pt] coordinates
{(-5,0) (0,0) (0,1) (5,1)};
\end{axis}
\end{tikzpicture}
\end{figure}
```

4.3 Sigmoid Function

```
\begin{figure}
    \centering
    \caption{Sigmoid Function.}
    \begin{tikzpicture}
    \label{fig.sigmoid}
    \begin{axis}[width=12cm,height=9cm,ylabel=$f(x)$,xlabel=$x$,ymin=0,ymax=1,xmin=-
5,xmax=5,
    legend pos= south east]
    \addplot[dashed,smooth] {1/(1+exp(-(0.5*x)))};
    \addlegendentry{$\alpha =0.5$}
    \addplot[black,smooth] {1/(1+exp(-x))};
    \addlegendentry{$\alpha =1.0$}
    \end{axis}
    \end{tikzpicture}
\end{figure}
```

4.3.1 Bipolar Sigmoid Function

```
\begin{figure}
    \centering
    \caption{Bipolar Sigmoid Function.}
    \begin{tikzpicture}
    \label{fig.bipolar.sigmoid}
    \begin{axis}[width=12cm,height=9cm,ylabel=$f(x)$,xlabel=$x$,ymin=-1,ymax=1,xmin=-
5,xmax=5,
    legend pos= south east]
    \addplot[black,smooth] {(1-exp(-x))/(1+exp(-x))};
    \addlegendentry{$\alpha=1.0$}
    \addplot[dashed,smooth] {(1-exp(-0.8*x))/(1+exp(-0.8*x))};
    \addlegendentry{$\alpha=0.8$}
    \addplot[dotted,smooth] {(1-exp(-5*x))/(1+exp(-5*x))};
    \addlegendentry{$\alpha=5$}
    \end{axis}
    \end{tikzpicture}
\end{figure}
```

4.4 Rectified Linear Unit (ReLU) (and Family of ReLU)

```
\begin{figure}
        \centering
        \caption{Rectified Linear Unit (ReLU) Family.}
        \begin{tikzpicture}
        \label{fig.relu}
        \begin{axis}[width=12cm,height=9cm,ylabel=$f(x)$,xlabel=$x$,ymin=-2,ymax=5,xmin=-5,xmax=5,
        legend pos= north west]
        \addlegendentry{$\alpha =0.00 \quad (ReLU)\quad$}
        \addplot[thick, mark=x,line width=2pt]
        plot coordinates {
                (-5,0)
                (0,0)
                (5,5)
        };
        \addlegendentry{$\alpha =0.01 \quad (LReLU)$}
        \addplot[dashed, mark=x,line width=1pt]
        plot coordinates {
                (-5,-0.5)
                (0,0)
                (5,5)
        };
        \addlegendentry{$\alpha = 0.25 \quad (PReLU)$}
        \addplot[dotted, mark=x,line width=1pt]
        plot coordinates {
                (-5,-2)
                (0,0)
                (5,5)
        };
        \addlegendentry{$\alpha = 0.80 \quad (RReLU)$}
        \addplot[thick, mark=x,line width=1pt]
        plot coordinates {
                (-5,-4)
                (0,0)
                (5,5)
        };
        \end{axis}
        \end{tikzpicture}
\end{figure}
```

4.5 Exponential Linear Unit (ELU) and Scaled Exponential Linear Unit (SELU)

\begin{figure}[!htb]
    \centering
    \caption{Exponential Linear Unit (ELU).}
    \begin{tikzpicture}
    \label{fig.elu}
    \begin{axis}[width=12cm,height=9cm,ylabel=$f(x)$,xlabel=x,ymin=-2,ymax=5,xmin=-5,xmax=5,
    legend pos= north west]
    \addlegendentry{$\alpha (e^{x}-1) \qquad (ELU)$}
    \addplot[thick, line width=1pt]
    plot coordinates {
        (0,0)
        (5,5)
    };
    \addplot[thick, smooth,line width=1pt]
    plot coordinates {
        (-5,-0.993262053)
        (-4,-0.981684361)
        (-3,-0.950212932)
        (-2,-0.864664717)
        (-1,-0.632120559)
        (0,0)}
    ;
    \addlegendentry{$\lambda=1.97 \quad (SELU)$}
    \addplot[dashed, smooth]
    plot coordinates {
        (-5,-1.956726244)
        (-4,-1.933918191)
        (-3,-1.871919476)
        (-2,-1.813389492)
        (-1,-1.45277501)
        (0,0)
        (5,9.5)}
    ;
    \end{axis}
    \end{tikzpicture}
\end{figure}

4.7 Odd Activation Function

\begin{figure}[!htb]
    \centering
    \caption{Odd Activation (Signum / Sign) Function.}
    \begin{tikzpicture}
    \label{fig.odd}
    \begin{axis}[width=12cm,height=9cm,ylabel=$f(x)$,xlabel=x,ymin=-2,ymax=2,xmin=-5,xmax=5]
    \addplot[thick,line width=2pt] coordinates
    {(-5,-1) (0,-1) (0,0) (0,1) (5,1)};
    \end{axis}
    \end{tikzpicture}
\end{figure}

4.9,10,11 Softsign, Elliott, and Tanh Functions

\begin{figure}[!htb]
    \centering
    \caption{Softsign, Hyperbolic Tangent and Elliot Function.}
    \begin{tikzpicture}
    \label{fig.softsign}
    \begin{axis}[width=12cm,height=9cm,ylabel=$f(x)$,xlabel=x,ymin=-2,ymax=2,xmin=-5,xmax=5]
    \addlegendentry{$Tanh$}
    \addplot[thick,smooth] {(2/(1+exp((-2)*x)))-1 };
    \addlegendentry{$Softsign$}
    \addplot[dashed,smooth] {x/(1+abs(x))))};
    \addlegendentry{$Elliott$}
    \addplot[dotted,thick,smooth] {((0.5*x)/(1+abs(x)))+0.5};
    \end{axis}
    \end{tikzpicture}
\end{figure}

4.11.1,2 Arc & Lecun's Hyperbolic Tangent Function

\begin{figure}[!htb]
    \centering
    \caption{Arc Tangent Function.}
    \begin{tikzpicture}
    \label{fig.arctan}
    \begin{axis}[width=12cm,height=9cm,ylabel=$f(x)$,xlabel=x,ymin=-2,ymax=2,xmin=-5,xmax=5]
    \addplot[black,smooth] {rad(atan(x))};
    \addlegendentry{$Arc Tangent$}
    \addplot[dashed,smooth] {1.7159*tan((2/3)*x)};

```
\addlegendentry{$Lecun's Tanh$}
    \end{axis}
    \end{tikzpicture}
\end{figure}
```

4.12 Complementary log-log Function

```
\begin{figure}[!h]
    \centering
    \caption{Complementary Log-Log Function.}
    \begin{tikzpicture}
    \label{fig.loglog}
    \begin{axis}[width=12cm,height=9cm,ylabel=$f(x)$,xlabel=$x$,ymin=0,ymax=1,xmin=-5,xmax=5]
    \addplot[black,smooth] {1-exp(-exp(x))};
    \end{axis}
    \end{tikzpicture}
\end{figure}
```

4.13 Softplus Function

```
\begin{figure}[!htb]
    \centering
    \caption{Softplus Function.}
    \begin{tikzpicture}
    \label{fig.softplus}
    \begin{axis}[width=12cm,height=9cm,ylabel=$f(x)$,xlabel=$x$,ymin=-2,ymax=2,xmin=-5,xmax=5]
    \addplot[black,smooth] {ln(1+exp(x))};
    \end{axis}
    \end{tikzpicture}
\end{figure}
```

4.14 Bent Identity Function

```
\begin{figure}[!htb]
    \centering
    \caption{Bent Identity Function.}
    \begin{tikzpicture}
    \label{fig.bent}
    \begin{axis}[width=12cm,height=9cm,ylabel=$f(x)$,xlabel=$x$,ymin=-2,ymax=2,xmin=-5,xmax=5]
    \addplot[black,smooth] {((sqrt((x^2)+1)-1)/2)+x};
    \end{axis}
    \end{tikzpicture}
\end{figure}
```

4.15 Soft Exponential Function

\begin{figure}[!htb]
        \centering
        \caption{Soft Exponential Function.}
        \begin{tikzpicture}
        \label{fig.softe}
        \begin{axis}[width=12cm,height=9cm,ylabel=$f(x)$,xlabel=x,ymin=-5,ymax=5,xmin=-5,xmax=5,
        legend pos= south east]
        \addlegendentry{$\alpha=-0.5$}
        \addplot[dotted, smooth] plot (\x, {-1*(ln(1-(-0.5*(\x-0.5))))*(-0.5^(-1))}); % alpha=-0.5 olsun.
        \addlegendentry{$\alpha=-0.2$}
        \addplot[densely dotted, smooth] plot (\x, {-1*(ln(1-(-0.2*(\x-0.2))))*(-0.2^(-1))}); % alpha=-0.2 olsun.
        \addlegendentry{$\alpha=\quad0$}
        \addplot[thick, smooth] plot (\x, {\x}); % Ortanca durum.
        \addlegendentry{$\alpha=0.2$}
        \addplot[loosely dashed, smooth] plot (\x, {((exp(0.2*\x)-1)*(0.2^(-1)))+0.2}); % alpha=0.2 olsun.
        \addlegendentry{$\alpha=0.5$}
        \addplot[densely dashed, smooth] plot (\x, {((exp(0.5*\x)-1)*(0.5^(-1)))+0.5}); % alpha=0.5 olsun.
        \end{axis}
        \end{tikzpicture}
\end{figure}

5.1.1 Sine and Cosine Function

\begin{figure}[!htb]
        \centering
        \caption{Sine and Cosine Wave Functions.}
        \begin{tikzpicture}
        \label{fig.sine}
        \begin{axis}[width=12cm,height=9cm,
        trig format plots=rad,
        axis lines = middle,
        enlargelimits,
        clip=false,
        ylabel=$f(x)$,xlabel=x,ymin=-1,ymax=1,xmin=-5.3,xmax=5.3,
        legend pos= south east]
        \addlegendentry{$sine$}
        \addplot[domain=-2*pi:2*pi,samples=200,thick, smooth] {sin(x)};
        \addlegendentry{$cosine$}
        \addplot[domain=-2*pi:2*pi,samples=200,dashed, smooth] {sin(x-2)};
        \draw[dotted,blue!40] (axis cs: 0.5*pi,1.1) -- (axis cs: 0.5*pi,0);

```
\draw[dotted,red!40] (axis cs: 0.5*pi+2,1.1) -- (axis cs: 0.5*pi+2,0);
\draw[dashed,olive,<->] (axis cs: 0.5*pi,1.05) -- node[above,text=black,font=\footnotesize]
{$\delta t$}(axis cs: 0.5*pi+2,1.05);
\end{axis}
\end{tikzpicture}
\end{figure}
```

5.1.2 Cardinal Sine Function

```
\begin{figure}[!htb]
    \caption{Sinc Function.}
    \begin{tikzpicture}
    \label{fig.sinc}
    \begin{axis}[width=12cm,height=9cm,
    trig format plots=rad,
    axis lines = middle,
    enlargelimits,
    clip=false,
    ylabel=$f(x)$,xlabel=$x$,ymin=-0.5,ymax=1,xmin=-10,xmax=10,
    legend pos= south east]
    \addlegendentry{$sinc$}
    \addplot[domain=-12:12,samples=200,thick, smooth] {sin(x)/x};
    \addlegendentry{$normalized$}
    \addplot[domain=-12:12,samples=200,dashed, smooth] {sin(pi*x)/(pi*x)};
    \end{axis}
    \end{tikzpicture}
\end{figure}
```

5.2.1 Gaussian (Normal) Distribution Function

```
\begin{figure}[!h]
    \caption{Periodic Gaussian Function.}
    \begin{flushright}
    \begin{tikzpicture}[scale=0.9]
    \label{fig.gaussian}
    \begin{axis}[
    domain=-12:12, samples=100,
    axis lines*=middle, xlabel=$x$, ylabel=$f(x)$,
    every axis y label/.style={at=(current axis.above origin),anchor=south},
    every axis x label/.style={at=(current axis.right of origin),anchor=west},
    height=5cm, width=15cm,
    enlargelimits=false, clip=false, axis on top,
    ]
    \addplot [thick,smooth] {gauss(0,1)};
```

```
\addplot [dashed] {gauss(8,1)};
\addplot [dashed] {gauss(-8,1)};
\end{axis}
\end{tikzpicture}
\end{flushright}
\end{figure}
```

5.2.2 Square Wave Function

```
\begin{figure}[!htb]
    \centering
    \caption{Square Wave Function.}
    \begin{tikzpicture}[scale=0.9]
    \label{fig.square.wave}
    \begin{axis}[ymin=-0.1,ymax=1.5,xmin=-1,xmax=7,
    width=12cm,
    height=9cm,
    x axis line style={-stealth},
    y axis line style={-stealth},
    xticklabels={},
    axis lines*=center,
    ytick={0.5,1},
    xlabel={Time $\rightarrow$},
    ylabel={Amplitude},
    xlabel near ticks,
    ylabel near ticks]
    \addplot+[black,thick,mark=none,const plot,line width=1.5pt]
    coordinates
    {(0,0) (1,1) (2,0) (3,1) (4,0) (5,1) (6,0) (7,1) (8,0)};
    \end{axis}
    \end{tikzpicture}
\end{figure}
```

5.2.3 Triangle Wave Function

```
\begin{figure}[!htb]
    \centering
    \caption{Triangle Wave Function.}
    \begin{tikzpicture}[scale=0.9]
    \label{fig.triangle.wave}
    \begin{axis}[ymin=-0.1,ymax=1.5,xmin=-1,xmax=7,
    width=12cm,
    height=9cm,
    x axis line style={-stealth},
```

```
y axis line style={-stealth},
xticklabels={},
axis lines*=center,
ytick={0.5,1},
xlabel={Time $\rightarrow$},
ylabel={Amplitude},
xlabel near ticks,
ylabel near ticks]
\addplot+[black,thick,mark=none,line width=1pt]
coordinates
{(0,0) (1,1) (2,0) (3,1) (4,0) (5,1) (6,0) (7,1) (8,0)};
\end{axis}
\end{tikzpicture}
\end{figure}
```

5.2.4 Sawtooth Wave Function

```
\begin{figure}[!htb]
\centering
\caption{Sawtooth Wave Function.}
\begin{tikzpicture}[scale=0.9]
\label{fig.sawtooth.wave}
\begin{axis}[ymin=-0.1,ymax=1.5,xmin=-1,xmax=7,
width=12cm,
height=9cm,
x axis line style={-stealth},
y axis line style={-stealth},
xticklabels={},
axis lines*=center,
ytick={0.5,1},
xlabel={Time $\rightarrow$},
ylabel={Amplitude},
xlabel near ticks,
ylabel near ticks]
\addplot+[black,thick,mark=none,line width=2pt]
coordinates
{(0,0) (2,1) (2,0) (4,1) (4,0) (6,1) (6,0) (8,1) (8,0)};
\end{axis}
\end{tikzpicture}
\end{figure}
```